Be a Successful
Green Builder

Be a Successful Green Builder

R . D O D G E W O O D S O N

New York Chicago San Francisco Lisbon London Madrid
Mexico City Milan New Delhi San Juan Seoul
Singapore Sydney Toronto

The McGraw·Hill Companies

Cataloging-in-Publication Data is on file with the Library of Congress.

McGraw-Hill books are available at special quantity discounts to use as premiums and sales promotions, or for use in corporate training programs. To contact a special sales representative, please visit the Contact Us page at www.mhprofessional.com.

1 2 3 4 5 6 7 8 9 0 DOC/DOC 0 1 4 3 2 1 0 9 8

ISBN 978-0-07-159261-1
MHID 0-07-159261-X

Sponsoring Editor
Joy Bramble Oehlkers
Production Supervisor
Pamela A. Pelton
Editing Supervisor
David Fogarty
Acquisitions Coordinator
Rebecca Behrens
Project Manager
Jacquie Wallace

Copy Editor
Wendy Lochner
Proofreader
Leona Woodson
Indexer
Leona Woodson
Art Director, Cover
Jeff Weeks
Composition
Lone Wolf Enterprises, Ltd.

The pages within this book were printed on acid-free paper containing 100% postconsumer fiber.

This book is dedicated to the memory of Travis.

Contents

CHAPTER   FIVE

Finding Financing the Easy Way

CHAPTER ELEVEN

Preparing Winning Bids   **161**

CHAPTER   FOURTEEN

Green Land Developing Could Double Your Income   221

Introduction

Sustainable building practices are the wave of the future. Going "green" is good for the environment and can be very good for your bank account. The demand for eco-friendly building practices is high. People are willing to pay more to waste less.

Whether you are a carpenter who wants to venture into being a builder or a traditional builder already doing business, you should give serious consideration to sustainable construction materials and practices. Going green is not as simple as using a few green products in the homes that you build. Look at the entire project. Integrate the various sustainable options to provide a full green package. This is the best way to earn your reputation as a green builder.

Before you take out new ads that label you as a green builder, you need to understand what responsibilities you are taking on. To be a real green builder, you are going to need to educate yourself from start to finish, and this book is the starting point.

The author, R. Dodge Woodson, is a household name in the construction industry. With over 30 years of experience as a builder, remodeler, and master plumber, Woodson knows what he is talking about. This is your chance to gain invaluable experience for a fraction of the cost of what a one-hour consultation would cost. If you are looking for a brighter future in the building business, this book should be your first step in building a solid foundation for your new career as a green builder.

Take a look at the table of contents. Flip through the pages and notice the tip boxes. The author has implemented his noted conversational tone and reader-friendly approach to make this book a pleasant trip rather than a grueling learning experience. The result is a very effective view of what it takes to become a successful green building contractor without the suffering often associated with learning. This is your fast track to getting in on the green boom.

The Business of Green Building

What is green building? Simply said, it is a method of construction that minimizes the effects on the environment. The growing trend to recycle, reuse, avoid abuse, and make the most of what you have with minimum impact on the future ecology of the planet has created a new opportunity for those involved in the construction industry.

Whether you are an established builder or just breaking into the business, green building can open some very profitable doors for you. Sustainable construction allows you the opportunity to increase your income while minimizing the impact on the earth. Sounds like a good deal to me.

What will give you the advantage to make more money? Many people are willing to pay more for sustainable products in their quest to improve the future for generations to come. Do you find it difficult to believe that people will pay more for environmentally friendly materials? The truth is that the green industry is bursting with growth and opportunity.

Green building involves more than wood. Plumbing fixtures that use a minimal amount of water to accomplish their task add to green building. Low-wattage and natural lighting contribute to a green home. Companies that pay more for the construction of their buildings may recover the added cost over years of operation. This is good for them and for you.

The green movement in construction begins with land development. Developers have learned to create green spaces in land planning. I started doing this back in the early 1980s. Sustainable building has grown and changed a great deal over the 30 years that I have been in the building business. Remodeling, building, plumbing, electrical work, land development, roofing, and many other elements of green building make it possible for you to add to or create your business.

There are lots of ways to enter the construction field: jumping right in and becoming a sole proprietor, joining up with some skilled trade associates to form a small company, or working with an established company while trying to pick up jobs on your own by moonlighting.

Lots of people know that plumbers, electricians, carpenters, painters, and other tradespeople work evenings and weekends to make extra cash. Many of the tradespeople I've known, myself included, have moonlighted at one time or another as a way of easing into a full-time business. Can a homebuilder get started by working nights and weekends? You bet, and this chapter will show you how.

Homebuilders, or residential contractors as they are referred to in the trade, originate from all walks of life. Many of them start out as carpenters, doing repair work or small remodeling jobs with an eye to becoming a full-scale homebuilder. I started out as a plumber and grew into remodeling. From there, I went on to build as many as 60 homes a year.

I've met builders who decided that their present occupation as a lawyer, real-estate broker, farmer, firefighter, or policeman was not for them, and so they left secure jobs for a shot at a new and exciting career. Almost anyone can transition from whatever they are currently doing to homebuilding. Making the transition is easier for some than it is for others.

If your background is in construction, you have an obvious advantage over someone who has never set foot on a construction site. While you may have never built a house, working in new houses, around other trades, gives a good idea of what goes on during the construction process. Field experience alone isn't enough to make someone a good builder—it sure helps, but there is a business side to building that also needs to be learned. How can you leave what you are doing for a living now and enjoy being a homebuilder?

Getting started as a builder is not easy; in addition to a small bankroll and some prior experience in construction, it will be very helpful if you have had some contacts with electrical, plumbing, and excavating subcontractors and equipment suppliers. And it would also be very helpful to have a relationship with a local bank, even if only to have a checking or savings account there. I'm sure that there are people with the financial resources to start a building business in a first-class manner. I never enjoyed this luxury. I had to start at the bottom and crawl up the mountain. At times it seemed as if the mountain was made of gravel, because every time I would get near the top, I would slide back down the hill. But I persevered and made it. I think you can, too.

After running my plumbing and remodeling business for some time, I set my sights a little higher, and homebuilding seemed like the next logical path to take. Since I didn't own a home, it made sense to try my hand at building by creating a new home for myself, and that is what I did. By building the home myself I was able to use my sweat equity as a down payment, so I essentially built the house without using any of my own money. I was even able to pocket a little cash as profit from the deal. That first house was the stepping stone that led to the top of my mountain of challenge. Along the way I also added to my credibility by establishing a relationship with some subcontractors and material suppliers, opened a few commercial accounts, and paid my bills on time.

When I say that I built my first house, I should qualify my terminology so that you will understand that I did not drive every nail personally. I acted as a general contractor. My wife and I did the plumbing, drywall work, painting, tile work, and some other odds and ends, but we used subcontractors for the balance of the work. The house wound up costing more than I had projected it would, and the drywall work wasn't all that great. Since building that house I learned a lot of lessons along the way, including hiring others to do drywall work. After building my own home in my spare time, I decided to sell it and start another new house.

For three years my wife and I built a new house each year. After building

PRO POINTER

Getting started as a builder is not easy; in addition to a small bankroll and some prior experience in construction, it will be very helpful if you have had some contacts with electrical, plumbing, and excavating subcontractors and equipment suppliers. And it would also be very helpful to have a relationship with a local bank, even if only to have a checking or savings account there.

our second new home, we started building houses on speculation. By the time we completed our third house, we were averaging about 12 new homes a year, in addition to continuing to work in our plumbing and remodeling business. When we really got rolling, we were doing 60 homes a year, running the plumbing business, doing a little selective remodeling, and operating real-estate-sales and property-management businesses. All this came about because of the income generated from homebuilding. The money made from our plumbing and remodeling operations paid our bills, and the moonlighting we did with spec houses gave us the cash to jump into the big leagues. I suspect that you could do something similar if you put your mind to it.

Basic Needs

What are the basic needs for becoming a part-time green builder? They are fewer than you might imagine. There are two different types of builders. One is the full-service general contractor—a person who hires all the tradespeople required for the job and uses few if any subcontractors. The other type is referred to as a "broker"—a general contractor who subcontracts all or most of the work to other "specialty" contractors (subcontractors).

If you operate as a "full service" contractor, you will have to hire your own carpenters, electricians, plumbers, framers, and roofer. This means having to meet large weekly payrolls in addition to finding qualified workers and enticing them to work for you.

If you operate as a "broker"-type general contractor, you don't need much in the way of trucks, tools, and equipment; the subcontractors you hire will provide their own. All you have to do is schedule and supervise their work. And since these subcontractors will generally bill you monthly, your cash flow will occur monthly instead of weekly. However, neither approach is as simple as it seems.

Since you will probably be working your day job when you get started, you will need an answering service or an answering machine to receive your

PRO POINTER

There are two different types of builders. One is the full-service general contractor-a person who hires all the tradespeople required for the job and uses few if any subcontractors. The other type is referred to as a "broker"—a general contractor who subcontracts all or most of the work to other "specialty" contractors (subcontractors).

phone messages. A license to build may be required in your region, and a business license will normally be required. You can work from home and meet your prospective customers in their homes. You should invest in liability insurance; your insurance agent can walk you through various types of policies. And there will be a need to advertise. On the whole, the financial requirements for becoming a part-time builder are minimal.

It is best for builders to have reserve capital to get past unanticipated financial problems, but if you're diligent in your work and a little lucky, you can get by with very little cash. If you bid jobs accurately and profitably, complete them on schedule, and check your books to ensure that your customers pay you on time, there may not be a need for a large reserve of cash—after all, you will have your regular employment to pay your routine bills and the small additional overhead costs of your building operation.

First Hurdles

There are two hurdles that need to be addressed as you begin to consider a career in construction: any requirements for licensing in the state in which you plan to operate and the lack of a track record and references.

Let's discuss the licensing issue. Many states require contractors to be licensed before they can operate their businesses. So it's a good idea to check

PRO POINTER

There are two hurdles that need to be addressed as you begin to consider a career in construction: any requirements for licensing in the state in which you plan to operate and the lack of a track record and references.

with your state government to determine if you need to apply for a license and, if so, what the qualifications are. They can vary considerably from state to state. I checked into the requirements for various states some time back. For example, in Maryland a license is required before you can work on home-improvement or remodeling projects. Keep in mind that it has been a while since I checked these facts, and you should always consult your current laws and regulations before taking part in any activity. Before you can get an application, you must pass an exam, and to qualify for the exam you must pass certain work and financial requirements. In Alabama, you just need four references, proof of insurance, and a net worth of $10,000. In Illinois, most construction contractors except for roofing contractors don't need to be licensed.

The second hurdle is the lack of a track record and references. Your customers will probably want the names of references, and they may even want to see examples of your work. When you are starting out, you can't provide references or work samples. This can be a difficult obstacle to clear, but there are some ways to work around the problem.

If you are building your first house on speculation, you won't have to worry about references or work samples. Potential buyers of the home will be able to see what they are getting. This makes your job a little easier. If you can't afford to take the risk of building on spec, you will have to be more creative in coming up with a way to get customers to accept you. If you have been doing repair or remodeling work in your off hours, you might use these customers as references.

The first houses I built and sold were homes that I built without having buyers lined up for them. Doing this created my references for me. But when I moved to Maine, I didn't have any references. And getting my first couple of contract homes to build was a bit more difficult. People would come into my office and talk about building, but they wanted to see tangible proof of my abilities, which I could not show them. So I had to come up with a plan.

To overcome the problem of not having a model or sample home to show people, I changed the types of advertisements I was running. The ads offered people a chance to have a new house built at a reduced cost by a builder who would pay them a monthly fee for the first year they owned the home in exchange for allowing their home to be shown periodically as a reference. I discounted my prices by 5 percent and offered buyers $150 a month for 12 months as a model-home fee. The strategy worked. Buyers came in, houses were built, and I was back up and running.

The discounted price and the $1200 per house for model fees cut my profit on the first two houses I built, but after those two homes were established as references, I didn't need to continue offering my special deal. By giving up a little in the beginning, I was able to get off to a fast start. You could try something similar, or maybe you will come up with a better idea. The point is that there are ways to get around the problem of not having references and sample homes.

Knowledge

How much knowledge of construction do you need to succeed as a homebuilder? The more you have, the better off you are. But you can get started with a basic level of knowledge and earn while you learn. As a general contractor, you don't have to do any of the physical work involved with construction. Your primary function once a job is underway is to schedule and supervise workers. It is obviously much easier to supervise people when you understand what they are doing and how the work should be done. But you don't have to be a drywall finisher to supervise drywall work. If a job looks good, you know it. When one looks bad, you can see it.

Code-enforcement officers from the local Building Department will be checking the work at various times—during foundation work, framing, and electrical and plumbing rough-ins to ensure that the work meets code. By sticking close to these officials during their routine inspections, you can pick up a lot of technical information.

In theory, you don't have to know much about construction to be a builder who subs all the work out to independent contractors, but as the general contractor it is you who will ultimately be responsible for the integrity of the work. Builders with the most knowledge of construction are generally much more successful than people who don't know about the homebuilding process.

If you have good organizational skills and manage people, budgets, and schedules well, you should be able to become a viable homebuilder. A lot of information can be obtained from watching various subcontractors perform their work. Ask around, and you'll find subcontractors willing to discuss their work and answer your questions. Reading books and trade magazines will add to your knowledge. Many specialized trade magazines are free; when you visit a subcontractor's office and see some of those magazines on the table, look for the tear-out forms that will allow you to apply for a subscription. There are numerous books and videos available for do-it-yourselfers that give step-by-step instructions for everything from plumbing to tile work. You can educate yourself by reading and watching tradespeople at work, but field experience is surely helpful.

The Dangers

One of the biggest dangers for rookie builders is their lack of experience in pricing homes. Even experienced carpenters often don't know how to estimate prices for complete homes. They are not accustomed to figuring in the cost of septic systems, sewer taps, floor coverings, or finish grading. How can you get the best estimates possible when you've never done one before?

PRO POINTER

One of the biggest dangers for rookie builders is their lack of experience in pricing homes. Even experienced carpenters often don't know how to estimate prices for complete homes. They are not accustomed to figuring in the cost of septic systems, sewer taps, floor coverings, or finish grading.

Take a set of blueprints to your supplier of building materials. Ask the manager to have someone assess your material needs and price them. Many suppliers will provide this service free of charge, but some won't. Circulate copies of the blueprints to every subcontractor that you will need. It is a good idea to get quotes from more than one subcontractor. Try to select two subs for each trade so that you set up a competitive situation. Have the subs give you prices for all the work they will be expected to do. While your subs and suppliers are working up their prices, you can start doing some homework of your own.

PRO POINTER

Once the appraiser has given you an idea of what the appraised value of your house would be, you can better gauge its market value. You can consult some pricing guides to determine what various phases of work will cost. These guides are available in most book stores, and they have multiplication factors that allow you to adjust the prices to coincide with those in your particular region.

Take a set of blueprints to a reputable real-estate appraiser. Ask the appraiser to work up either an opinion of value or a full-blown appraisal. This will cost you some money, but it will be well worth it.

Once the appraiser has given you an idea of what the appraised value of your house would be, you can better gauge its market value. You can consult some pricing guides to determine what various phases of work will cost. These guides are available in most book stores, and they have multiplication factors that allow you to adjust the prices to coincide with those in your particular region.

When you get your prices back from the suppliers and subcontractors, you can compare them with the numbers you came up with from the pricing guides. You can also look at the difference between the bid prices of your subs and suppliers and the finished appraisal figure. The spread between the bids and the market value represents your potential profit. It should normally relate to about a 10- to 20-percent gross profit. A 15-percent profit might be an average, but the amount varies with economic conditions and the quotes you received from subcontractors and suppliers. We are going to talk more about pricing and estimating later in the book, but the procedures we have just discussed are the basics.

The Internet provides another source for estimating services. The Marshall & Swift website is just one of those sources. There are several cost guides available, such as Sweets Repair and Remodel Cost Guide, published by the McGraw-Hill Companies, and Means Residential Cost Data, published by the RS Means Company.

How Many Houses Can You Build?

How many houses can you construct in a year as a moonlighting builder? The answer to this question will vary with experience levels, sales opportunities, lot availability, individual time commitments, and so forth. You should certainly be able to build at least two houses and probably four. Since you are using subcontractors, you can have multiple houses under construction at the same time. By phasing the start of multiple housing projects, you can spread out the cash demands over a longer period of time.

I used to do about 12 houses a year as a part-time venture, but I had help from my wife. A goal of six houses a year seems as if it would be realistic for someone with a construction background and a good stable of subcontractors. Once you get into the building game, you can assess your time needs and adjust your goal as needed.

How Much Money Can You Make?

How much money can you make as a part-time builder? It all depends on the type and size of the houses you are building and how well you manage your production schedule and financial budget. If you are building $200,000 homes and your gross profit is 20 percent, you are making $40,000 on each house you build. Four houses a year at

PRO POINTER

How much money can you make as a part-time builder? It all depends on the type and size of the houses you are building and how well you manage your production schedule and financial budget.

this rate would be $160,000. Can you really do this? It can be done, but you will probably make less money per house until you gain more experience. But even at $30,000 a house, you're still making $120,000, and that's not bad for a part-time job.

If you want to look at the worst-case scenario, assume that you will make a 10-percent profit. That's $20,000 a house, or $80,000 a year. There really is a lot of money to be made as a builder once you work the bugs out of your system. If you can survive the first few houses you build and gain a reputation as an honest, quality builder, there is a very good chance that you can enjoy a long and profitable career.

The figures that I have just given you are based on you being a general contractor and not doing the physical work yourself. If you are your own lead carpenter, you should make your wages and still make your full percentage. Any of the work that you can do yourself will add to your bottom line. But remember, the real money is in volume. If you get too bogged down driving nails, your volume as a general contractor will suffer.

When you factor in the potential boost that being a green builder can give you, your odds for success are even better. Playing up your dedication to sustainable construction will increase your base of potential buyers. You can build traditional homes and green homes. Or you can specialize in green building and hone your reputation within that group of customers. The key is to have customers, and sustainable building can provide you with more income opportunities.

Setting Up Your Business Structure

Getting your feet wet in the building business without getting in over your head takes planning. Don't just decide one day to be a homebuilder, run an ad in the local newspaper, and wait for the phone to ring. This is a big step in your career and not one to take lightly; a lot of planning and thinking needs to be done before that first nail is driven. The financial rewards in this business can be substantial, but remember that where there is the opportunity for reward there is also the potential to lose money. To avoid failure, you must have some solid plans.

The first thing you need to do is make sure that you are ready to assume the responsibilities that go with being a builder. Do you have enough general knowledge to perform the functions of a general contractor? If you don't, start reading, attending classes, or working on some construction sites. Gain as much experience and knowledge as you can before you offer your services as a builder to the public.

If you are a seasoned builder but are just stepping into the field of green building, you will have plenty of research to do. Get familiar with the types of green products available to you. Find out the procedures for sustainable building. Join some builder organizations where sustainable building is a topic of conversation. Get online and join some forums and newsgroups. Immerse yourself in the learning process and become proficient in your knowledge of green building.

There are many ways to prepare for becoming a general contractor. Read every book you can find on building and related trades, and read books written for homeowners and do-it-yourselfers. Seek out titles that have been written for professionals, such as those published by McGraw-Hill. Absorb the wealth of knowledge provided by seasoned professionals that can be found in these books. Talk to people in the business. You'll find many builders who are willing to share some of their experience with you and point out some helpful tips to get you started as well as some pitfalls to avoid.

You can also attend classes that pertain to various trades and homebuilding. Some places, such as the area where I live, offer support workshops for people hoping to build their own homes. Look into the possibilities of attending workshops or vocational classes if you feel you need more training than you can get from a book. Local community colleges frequently offer construction-related courses such as basic carpentry skills, blueprint reading, and even project management, and the classes are generally held in the evening and are relatively inexpensive.

PRO POINTER

There are many ways to prepare for becoming a general contractor. Read every book you can find on building and related trades, and read books written for homeowners and do-it-yourselfers. Seek out titles that have been written for professionals, such as those published by McGraw-Hill. Absorb the wealth of knowledge provided by seasoned professionals that can be found in these books. Talk to people in the business. You'll find many builders who are willing to share some of their experience with you and point out some helpful tips to get you started as well as some pitfalls to avoid.

Videos have become extremely popular, and there are many available that show how to perform certain tasks, such as hanging cabinets or installing plumbing. Your local library or video store may have some of these learning tools on hand. If not, I'm sure they can help you order titles that will boost your skill level.

Go to some local residential building sites and check out what's going on. Make note of any subcontractors working there or names of companies delivering materials or equipment. This will help you become more familiar with the local construction community. Walk through some of the homes under construction and observe the types of materials and products being used. Watch some of the work being performed to gain more knowledge of the trades involved.

Trade shows can be an excellent source of product knowledge. Given all the publicity generated by and for green building, it should be easy to find trade shows, open houses, seminars, and other gathering places to go in search of a higher understanding of sustainable building.

The business side of homebuilding is critically important to your success. If you are not comfortable with your office skills, such as basic management and accounting principles, once again, investigate the business courses offered by your local community college. Good administrative skills will make your entry into the building business much easier.

Now, assuming that you feel ready to become a builder, you must find a way to tap into this lucrative market. If you have some money and good credit, you could start by building a spec house. This can be extremely risky, however, and I wouldn't recommend it as a starting point, even though it is how I got started. The safest way to break into building is with houses for which you already have buyers.

Which Type of House Should You Build First?

Which type of house should you build first? There are advantages to building some styles of houses over others. As your first house, you would not want to get involved in the construction of a complex design on a site that may have soil or drainage problems. Remember, you want to be able to use that first homebuyer as a reference on your next project. A ranch-style house is the easiest to build. Straight-up two-story homes give buyers the most square footage for their dollar. Cape Cods are very popular with people who are looking to build their first home on a budget. Elaborate designs make a builder's job tougher. Since you are going green, incorporate as many green elements into your first home as possible.

If you are planning to work with experienced subcontractors, your choice of type and style of your first home is broader, but logic dictates that simple designs are faster and easier to build. It is reasonable that you should start with a house that can be built quickly, so you can generate a cash flow and produce a profit as soon as possible. Keeping your starting designs simple will make it

PRO POINTER

If you are planning to work with experienced subcontractors, your choice of type and style of your first home is broader, but logic dictates that simple designs are faster and easier to build. It is reasonable that you should start with a house that can be built quickly, so you can generate a cash flow and produce a profit as soon as possible. Keeping your starting designs simple will make it easier to estimate material and labor needs, with less risk of cost overruns. With all of this in mind, I would recommend three sets of house plans to start off with: a ranch, a cape, and a two-story.

easier to estimate material and labor needs, with less risk of cost overruns. With all of this in mind, I would recommend three sets of house plans to start off with: a ranch, a cape, and a two-story.

You may be wondering why you should come up with specific house plans if you are going to be doing custom building. This is a fair question. You will ultimately be building to explicit specifications supplied by your customer, but you need a base design or designs to use in your advertising program. If you simply run an ad that says you are a homebuilder and that you are open for business, you may not get a lot of buyer activity. But if you run a picture of a particular house, list the features of the home, and include a price, your phone should start to ring. Obviously, to do this, you need to have a set of building plans and specifications to work with as a starting point.

> **PRO POINTER**
>
> Part of the decision as to which style of home to choose for your ads should be based on local conditions. New England-style Cape Cod homes may not fit into an area where two-story brick homes, ranch-style homes, or Spanish Mission homes are being built and sold. Economic factors come into play, and the price range for the new home should fit the area where it is to be built.

Part of the decision as to which style of home to choose for your ads should be based on local conditions. New England-style Cape Cod homes may not fit into an area where two-story brick homes, ranch-style homes, or Spanish Mission homes are being built and sold. Economic factors come into play, and the price range for the new home should fit the area where it is to be built. You're going to be competing with established builders, so you need an edge or an angle to work with. Gaining this advantage is often easier than you might imagine. Going green is a very good ace to play in the competition game. Not all builders have taken this path, so you could have a strong advantage by playing up sustainable building.

Choosing the Right House Plan

Choosing the right house plan is not something that should be done based on personal preference. You'll need to do some research; just ride around and see which types of houses are popular in the area where you are planning to build. Are a lot of split-foyers being built? How many newer homes have only one level of living space?

FIGURE 2-1 It is best to begin with simple designs which are faster and easier to build, such as the one pictured here (courtesy of ECO-Block, LLC).

Do most new homes have attached garages? Are front porches fashionable? Looking for these types of design features will bring you up to speed quickly on what the public wants. If you see that only one out of eight houses is a one-story ranch design, you can shuffle your ranch plans to the bottom of your pile.

Don't try to buck trends your first time out; your chances of success are much better if you follow the lead of your competitors. If colonial two-story homes are abundant in your area, look for a nice set of plans for a colonial. Go with the flow, but with a twist. What is the twist? Green building is your angle.

Your Edge

What is your edge as the new builder on the block? It's something you have to create. By going green, you are starting off on the right foot. It might be low price or an outstanding design or superior workmanship at an affordable price. Something as obvious as marketing and advertising could be what sets you above your competition. The edge can be almost anything, but you need it to survive and prosper. If you are just a carbon copy of all the other builders, you will be at a disadvantage. Finding what will work best for you is a personal thing, but I can give you some ideas.

Price

Price is a factor that many businesses use as a lever. Trying to beat the prices of your competitors would not be my first recommendation. If you become known as the cheap builder or the discount builder, you will have trouble moving up to higher-priced homes. But getting a reputation as a value-conscious builder and a quality-oriented builder is a different story. Everyone enjoys receiving good value for their money, but some people view discounted products as damaged merchandise.

To create the aura of a value-based builder, you have to make your homes different. Set yourself apart by pointing out the benefits of building a green home or office building. You can put your two-story colonial up against your competitors, but you may have to create some subtle differences in the plan. Your goal is to make customers compare apples to oranges rather than apples to apples. This way your homes don't appear to be a cheaper version of your competitor's. If your competition has a laundry room built onto the side of a home, as many colonials do, you might consider putting the laundry hookup in a basement or closet to eliminate the cost of the attached room. With these savings you might even be able to include a washer and dryer in that closet or basement laundry area. This design change may affect the exterior appearance of the home, but it also lowers the cost in a way that can't be construed as a discount, and if you decide to include the washer and dryer, it might even be looked upon as "added" value!

PRO POINTER

Stress the importance of sustainable housing for the generations to come. Few parents buying a house will not consider the future of their children. Keep the homes organic and kid-friendly if you are building for new and growing young families. There are countless ways to promote the benefits of green building to give you a decided edge.

Stress the importance of sustainable housing for the generations to come. Few parents buying a house will not consider the future of their children. Keep the homes organic and kid-friendly if you are building for new and growing young families. There are countless ways to promote the benefits of green building to give you a decided edge.

Identify Your Customer Base

Before you make your blueprint selection, you must identify with your customer base. Will you be dealing with first-time buyers and starter homes, or will you be working with more affluent buyers? There is less money to be made on a per-deal basis with starter homes, but these entry-level houses are a good place to begin your building business. There are several reasons for this. First-time buyers are often ignored by the larger, more successful contractors. This opens the market up for you. Secondly, first-time buyers are not usually too choosy when making a commitment to purchase a house. First-time homeowners are excited just to have a place to call their own and will be easier to please because they have no preconceived ideas about what they are looking for. Another big advantage of first-time buyers is that they are not encumbered with a house to sell. This makes it possible for you to make quick sales that are not contingent on the sale of other properties.

Second-time homebuyers often have to liquidate their existing homes before they can make a building commitment. Having owned a home previously, these buyers are frequently more selective than first-time buyers; they may look for features they didn't have in their old home but must have in their new one. The price of houses purchased by move-up buyers is more than that of the starter homes for first-time buyers, but the difference can be made up in volume and quick turn-over with first homes.

I can't tell you absolutely what will be the best choice for you. But I can tell

PRO POINTER

Second-time homebuyers often have to liquidate their existing homes before they can make a building commitment. Having owned a home previously, these buyers are frequently more selective than first-time buyers; they may look for features they didn't have in their old home but must have in their new one. The price of houses purchased by move-up buyers is more than that of the starter homes for first-time buyers, but the difference can be made up in volume and quick turn-over with first homes.

you that I have catered to first-time buyers as both a builder and a broker, and I've done well in doing so. If I were you, I'd give serious consideration to focusing on first-time buyers.

Financing

Financing programs can create an edge for you. Not that you have to provide the financing yourself; if you establish relationships with various lenders and can advertise some of their financing plans that are available for the homes you build, customers should respond favorably. Financing is usually the key to whether or not a building deal flies. Your competitors may have access to the same loans, but if you're the one making people aware of the available programs, you're likely to win the jobs.

Bringing It All Together

Bringing it all together to offer the public a fast, easy package is a sure way to success. Homebuyers are usually excited but often naive. Most buyers will respond to advertising, and almost anyone will listen to a sincere presentation from a caring, knowledgeable builder. You don't have to be the biggest builder in town to capture your share of the market. But, you do have to be professional and persistent and have a reputation for honesty and integrity.

So many options exist for potential homebuilders that it would take several chapters of a book to list most of them. Let me give you an example of how you might bring your building business into reality with minimal risk. You can use the example as a template for other types of approaches.

Assume that you have done your homework and decided to go after first-time homebuyers by building Cape Cods with unfinished upstairs areas and small ranch-style homes. You will be competing against modular homes and

PRO POINTER

Bringing it all together to offer the public a fast, easy package is a sure way to success. Homebuyers are usually excited but often naive. Most buyers will respond to advertising, and almost anyone will listen to a sincere presentation from a caring, knowledgeable builder. You don't have to be the biggest builder in town to capture your share of the market. But, you do have to be professional and persistent and have a reputation for honesty and integrity.

local established builders. With the money you have available, you might plan to do a little advertising on cable television, some print advertising in the local paper, a small direct-mail campaign, or some combination of the above. Now that you've got your house styles selected and your construction costs calculated, it's time to launch your attack.

Let's say you have decided to run a few ads in the local newspaper and on cable television. Once you get some name recognition, you may decide to use a direct-mail strategy. There are companies that sell mailing lists depending upon geographic area, economic level, family size, renters, home-

PRO POINTER

Assume that you have done your homework and decided to go after first-time homebuyers by building Cape Cods with unfinished upstairs areas and small ranch-style homes. You will be competing against modular homes and local established builders. With the money you have available, you might plan to do a little advertising on cable television, some print advertising in the local paper, a small direct-mail campaign, or some combination of the above.

owners, and so on. The mailing list you decide to use should consist of people who live in rental property and who have adequate income levels to afford the houses you plan to build.

As a format and theme for your new company, let's say that you have decided to emphasize certain points in your offering that are focused on first-time buyers. You indicate that easy financing is available, and your house designs have been chosen with young families in mind. Emphasize sustainability, quality construction, and energy efficiency as cornerstones of your business. Flexibility and freedom of design are major selling points. You can guarantee a quick turnaround on building that will put customers in their new homes in less than 90 days from the date that trees are cleared and ground is broken. You're offering a 10-year home warranty, endorsed by a major building association. All your features and benefits are listed in your advertisement. From start to finish, you will make homeownership easy and enjoyable; that's your motto and that is your goal.

When you start running your ads, you may be amazed at the response. Why are so many calling you, the new builder in town, when they could be calling established professionals? They're doing it because you identified a need, filled it, and make the public aware of what you were doing. I've done this type of thing time and time again.

Established builders may become complacent because they remain profitable and get enough work from word-of-mouth referrals that they don't have to actively look for new jobs. When first-time buyers approach some builders and real-estate brokers, they are not treated as potentially valued customers; I've heard this complaint from buyers on countless occasions. These first-time buyers feel as if the builders and brokers don't want their business, and this group of buyers is a prime target for your approach. When you are willing to talk to them as equals and offer advice on things to look for in a new home, buyers will flock to you and spread the word to their friends about how great you are to deal with.

Test the Waters

When you are ready to get into building, you have to test the waters to develop potential markets. If money is no object (and there aren't many of those builders around), you can test the market by investing in newspaper or magazine advertising. Research is the key. Talking with competitive builders, real-estate brokers, and appraisers is a fast way to get some inside information. Visiting nearby residential developments under construction will give you an idea of what is selling, both style- and price-wise. Looking through comparable-sales books distributed by multiple-listing services to brokers is also a valid way to see what's selling and how much it's selling for. Comp books also let a person know how long a property was on the market before it sold.

You will have to carve your own niche in the world of building, and there are many of them, as we will discuss in later chapters. Mine has been first-time buyers. Some builders specialize in expansive, expensive homes. The profit builders of big homes make on one job might take me four jobs to make, but I have the volume. What worked for me might not work for you, but I really think that the plan I have presented in this chapter represents a good starting place for you. As we proceed, you will learn more about defining your future.

20 Key Mistakes to Avoid When Becoming a Building Contractor

Over the years, I've done a lot of building, and I've come to know more builders than I can count. My work as a consultant to other contractors has exposed me to a host of problems. I've suffered a multitude of difficulties personally, and my work as a consultant has shown me mistakes of others that I haven't yet made. The combination of my experience as both a builder and a consultant has given me a great deal of respect for the building business. It doesn't take but one slip-up to put a builder out of business. Someone once said that smart people learn from experience, but wise people learn from the experience of others—so hopefully you will learn from others' mistakes.

Builders just starting out in business will probably make mistakes that are likely to be financially fatal. But experienced builders often fall into traps of their own, and I'm not sure that the rookies are at the highest risk. When people start doing something new, they are often careful with every step they take. Once they feel like they are in control, they tend to lighten up on their caution. This is usually when trouble strikes. So even if you've been building houses for 15 years, you can mess up.

Everyone makes mistakes. My mother used to tell me that the first time I did something wrong it was experience and that the second time I did the same thing wrong it was a mistake. This makes sense, but unfortunately business owners sometimes only get one chance to do their job right. This can be especially true for homebuilding, where large sums of money are at risk.

The amount of money at stake when houses are being built can be overwhelming and more than the average person could reasonably be expected to pay back if something disastrous happened. I remember a time when I was leveraged out for over four million dollars in construction loans. If something had happened to me at that time of my life, there would have been no way that I could have ever paid the money back by working a normal job.

There are many mistakes made by contractors and most of the problems can be avoided, but what do you look for? Experience is often the only protection we have against mistakes. How do you survive in a tough business long enough to get the experience you need? It is a difficult situation.

You have taken a major step in the right direction by reading this book. By learning from my experience and mistakes, you will be better prepared to avoid your own. Seeing the danger signs in time to react before you are in too deep is paramount to your success as a builder. Hopefully, you will gain enough knowledge from this book and other sources to make your way to the top of your profession.

I've thought over all the mistakes that I can remember either making or observed being made. My list is a long one. Many of the problems, however, are not so deadly as to sink your business. So I will share with you 20 of the all-time mistakes that I'm aware of. If you study the topics we are about to discuss, you will see how you might avoid mistakes that builders frequently make.

It Takes More Money Than You Think

It takes more money to become a full-time builder than you might think. This is one of the first mistakes builders often make. If you sit down and run numbers on what your startup costs will be, the numbers will probably look manageable. The hard-money expense of opening a building business isn't very high. It's the hidden expenses that will put you out of business before you get started.

A lot of people either don't know about or don't think about many of the expenses they will incur when going into business for themselves. For example, you will still have all your personal bills to pay after you quit your job

PRO POINTER

It takes more money to become a full-time builder than you might think. This is one of the first mistakes builders often make. If you sit down and run numbers on what your startup costs will be, the numbers will probably look manageable. The hard-money expense of opening a building business isn't very high. It's the hidden expenses that will put you out of business before you get started.

and while you are waiting to draw your first income from your business. That steady paycheck won't be there for you each week. If you don't have enough money saved to support yourself for several months, your time as a professional builder could be very short. Income is not the only issue. If you presently have health-insurance benefits supplied by your employer, you might not think to factor the cost of insurance into your new business overhead projections. Forgetting this fact could put you at the unhappy end of a surprise when you realize that hundreds of extra dollars will be needed each month for business insurance premiums.

Failing to prepare properly when going into business is one of the major reasons why so many businesses never see their one-year anniversary. Take the time to plan out all aspects of your business before you cut your ties with an employer.

Avoiding Heavy Overhead Expenses

Avoiding heavy overhead expenses is something that every business owner should strive to do. Most builders know that high overhead expenses can be dangerous, but many of them dive right in anyway. If you rush out and rent a lavish office, buy a new truck, lease a heavy-duty copy machine, and use up your available credit on things that are not truly needed, you may find yourself in a cash bind. Terminating long leases and installment debt can be very difficult and expensive. Not being able to lower or eliminate your overhead can put you out of business.

PRO POINTER

Avoiding heavy overhead expenses is something that every business owner should strive to do. Most builders know that high overhead expenses can be dangerous, but many of them dive right in anyway. If you rush out and rent a lavish office, buy a new truck, lease a heavy-duty copy machine, and use up your available credit on things that are not truly needed, you may find yourself in a cash bind.

It's easy to get caught up in the excitement of owning your own business. Part of the fun is getting to buy things that you want. But you must be reasonable about your purchases and expense commitments. A full-time assistant is something that you probably don't need right away. There are dozens if not hundreds of ways to waste your money before you have even earned it. Move cautiously, and don't put yourself in a financial box that you can't get out of.

Too Cautious

There is such a thing as being too cautious. If you tend to be a money miser, you might find that you're not cut out to be an entrepreneur. To make a new business work, you have to take risks and spend money. I know I just finished telling you to be cautious, but don't let fear grasp and immobilize you.

I've seen contractors start a business and refuse to pay one penny for advertising. How they expect to get business if no one knows that they are in business is beyond me. My personal experience and mistakes have made it abundantly clear to me that a person can hurt him- or herself by trying to save money without thinking of the consequences. I once terminated a big, expensive ad I had in the phone book to save money. Well, I didn't have to pay the ad expense any longer, but my business dropped off dramatically. I lost much more money in reduced business than I saved. Balance your spending to achieve the best results.

Select Your Subcontractors Carefully

Even though your subs are independent contractors, they have a lot to do with your business image. Sloppy workmanship by your subcontractors will reflect on you; remember, you are the general contractor and responsible for everything that happens on your job site. If your subs are rude to your customers, you'll take the heat. Many builders are attached to subcontractors who offer the lowest prices. This can be a major mistake. Don't assume that price is the only factor when selecting a subcontractor. Screen your subcontractors thoroughly before you put them on a job. It only takes a few bad jobs for word to get around that you are a builder to avoid.

PRO POINTER

Even though your subs are independent contractors, they have a lot to do with your business image. Sloppy workmanship by your subcontractors will reflect on you; remember, you are the general contractor and responsible for everything that happens on your job site. If your subs are rude to your customers, you'll take the heat.

Set Up a Line of Credit

Set up a line of credit with your banker before you need it. People often wait until they need money to apply for a loan. This is usually the worst time to ask for money from a bank. My experience with bankers has shown that they are much more likely to approve loans when you don't need them rather than when you do need money.

Most builders run into cash-flow crunches from time to time. Having a line of credit established to tide you over the rough spots until your next draw or payment comes in can make a lot of difference to both your building operation and your credit report. If you delay paying your suppliers for a couple of months until money comes in, the suppliers may not go out of their way to do business with you in the future. Sometimes houses don't close on schedule, and a builder's money can be held up for several weeks or even months. Unless you have sufficient funds in a reserve account, a line of credit in anticipation of problems will save the day.

PRO POINTER

Most builders run into cash-flow crunches from time to time. Having a line of credit established to tide you over the rough spots until your next draw or payment comes in can make a lot of difference to both your building operation and your credit report.

Get It in Writing

When you are in business, it's always best to get all the details in writing. For a builder, it's especially important to have contracts with both customers and subcontractors. As a builder you will deal with lots of contracts—why do you think they call us contractors? You should always insist on a letter of commitment that shows that a loan has been approved for your customer before you start construction. Some builders start work on custom homes as soon as a customer signs a contract. You can do this, but it's very risky. If your customer is denied a loan, you

PRO POINTER

When you are in business, it's always best to get all the details in writing. For a builder, it's especially important to have contracts with both customers and subcontractors. As a builder you will deal with lots of contracts—why do you think they call us contractors? You should always insist on a letter of commitment that shows that a loan has been approved for your customer before you start construction.

may never get paid for any of the work or material you have supplied. I hate to say it, but don't take your customer's word for the fact that a loan has been approved. I've never known a lender who didn't issue a letter of commitment once a loan was approved. You should have a copy of the commitment letter in your job file. Even if you are working only as a construction manager, you should require your customer to sign a letter of engagement.

Stay Away from Time-and-Material Prices

Stay away from time-and-material (T-&-M) work when dealing with your subcontractors; it is like giving them an open checkbook. It's possible to save money with T-&-M deals, but you can also lose a lot of money. The lure of saving some money is a strong one, but be aware that your attempt to save could backfire and cost you plenty.

I built a house for a customer a couple of years ago and used a site contractor on a T-&-M basis. The contractor was recommended to me by one of my carpenters. After meeting with the contractor, I found that he was just going into business for himself and wasn't sure how to give me a firm price. As experienced as I was, I agreed to a T-&-M basis. I did this to save money and to help the guy out. The deal backfired on me. I'd received quotes from other contractors, and the T-&-M price wound up being much higher than any of the quotes. Lump-sum pricing is sometimes more expensive, but it's always a safe bet.

Check Zoning Regulations

Before you chop down that first tree or dig your first hole, check the local zoning regulations. I know many occasions when seasoned builders have gotten into big trouble with zoning problems. A commercial building erected in Virginia over a setback line had to be moved. A local builder here in Maine recently built a house, and part of the foundation was on someone else's land. A carpenter who previously worked for me went into business for himself and placed a well on a building lot in a position that made it impossible

PRO POINTER

Zoning regulations are easy to check into. If you go to your local code-enforcement or zoning office, you can find out exactly what you need to know about setbacks, easements, and other restrictions. If you build a house in a restricted or prohibited space, you're likely to have a major lawsuit filed against you. This can be avoided with some simple investigative work.

to install a septic system without getting an easement from the adjoining land owner. I could go on and on with these types of stories.

Zoning regulations are easy to check into. If you go to your local code-enforcement or zoning office, you can find out exactly what you need to know about setbacks, easements, and other restrictions. If you build a house in a restricted or prohibited space, you're likely to have a major lawsuit filed against you. This can be avoided with some simple investigative work.

Check Covenants and Restrictions

Many subdivisions have covenants (binding agreements) and restrictions that protect the integrity of the development. Customers who buy a piece of land and want to have a custom home built may not be aware that these restrictions exist. Some builders buy lots without ever asking what restrictions apply. Some examples of common restrictions include a minimum amount of square footage, a prohibition on the use of some types of siding and roofing, and even requirements for the style or type of house that can be built. Don't build a house and then find out that the siding you used has to be torn off and replaced. Even more importantly, don't build a ranch-style house and then discover that the subdivision only allows two-story homes. This can happen.

PRO POINTER

Many subdivisions have covenants (binding agreements) and restrictions that protect the integrity of the development. Customers who buy a piece of land and want to have a custom home built may not be aware that these restrictions exist. Some builders buy lots without ever asking what restrictions apply. Some examples of common restrictions include a minimum amount of square footage, a prohibition on the use of some types of siding and roofing, and even requirements for the style or type of house that can be built.

You can check for covenants and restrictions yourself by reading a copy of the deed to a piece of property. If you're buying a lot, have your attorney check the prohibitions prior to making a final purchase commitment. When a customer comes to you with a lot he or she already owns, ask to see a copy of the deed. Protect yourself. Nobody is going to do it for you.

Make Sure You Have Enough Insurance

Some states require builders to carry a minimum amount of insurance, and others don't. Whether your local laws require insurance or not, you should get it. At the very least, you need liability insurance. There are many other types of coverage that may be needed, such as worker's compensation insurance. Talk to your insurance agent for advice on exactly which types of coverage you should have.

If you build without the proper property and general liability insurance, like a few builders I know, you are sitting on a time bomb. One lawsuit is all it takes to ruin your life. You could lose your business, your home, and your future. Insurance premiums may seem like an unnecessary expense to pay, but if you ever need the coverage, you'll be thankful that you have it.

Don't Give Inaccurate Quotes

Inaccurate quotes can have you working for nothing. Most builders commit to build a house for a lump sum or flat fee. If the price you quoted is too low, too bad for you. While you may not bid a job so low that you actually lose money, it's very possible that you will give away some percentage of your profit due to mistakes in your pricing.

Building is a business that often runs in cycles. You might go for months with no work in sight and suddenly get flooded with requests for quotes. This is a dangerous situation. You've been sitting around with nothing to do and wondering how you will ever pay the bills. When you finally get a chance to bid a job, you might tend to submit a low bid to make sure you get it. This may be bad enough, but it could get worse. When bid requests are piling up on your desk, you might rush through them, hoping to make up for lost time. You can make mental errors under these conditions and lose a lot of money.

If you make mistakes in your take-offs or estimates, you may not know it until a house is nearing completion. By then, it's usually too late to do anything other than reconcile yourself to losing

PRO POINTER

Take your time and make sure your estimates are right. Double-check everything you do. If you have someone who can review your numbers, such as one of your field supervisors, do it. Fresh eyes often catch omissions. Once you commit to a price in a contract, you will be stuck with it.

money. Many builders, myself included, have lost money by making mistakes when quoting jobs, and I've lost lots of sleep over the experience. Take your time and make sure your estimates are right. Double-check everything you do. If you have someone who can review your numbers, such as one of your field supervisors, do it. Fresh eyes often catch omissions. Once you commit to a price in a contract, you will be stuck with it.

Inspect Your Jobs Frequently

Inspect your jobs frequently. Some builders are reluctant to go out into the field on a regular basis to check on schedules and quality of work. They're either too busy, reluctant to run into the customers, or just too lazy. You need to visit your job sites often. I try to get to every job at least once a day. When I was doing volume building, I had two full-time supervisors checking every job twice a day.

PRO POINTER

As the general contractor, you should stay on top of your jobs as best you can. Failure to inspect jobs results in poor quality control. When this happens, a builder's reputation can be tarnished. If you get a bad reputation in the building business, you might as well look for some other type of work.

A lot can happen in just one day. If you skip an inspection, your customer may know more about the condition of a house than you do. It's very embarrassing to have a customer come to your office making a complaint about a condition that you are not aware of. As the general contractor, you should stay on top of your jobs as best you can. Failure to inspect jobs results in poor quality control. When this happens, a builder's reputation can be tarnished. If you get a bad reputation in the building business, you might as well look for some other type of work.

Maintain Customer Relations

Customer relations are a critical part of a building business. Building a house is a very emotional experience for the homebuyer and sometimes he or she will become irrational. You have to be willing to talk with your customers, and you may have to settle them down from time to time. If you don't pay attention to your customers, you won't get the coveted referral business that so many builders thrive on. It's essential that you keep your customers happy.

Use Change Orders

If your customers ask for added services or changes in contract obligations, use change orders to document the requested change or additional service you are asked to provide. Otherwise, you could be left holding the bag on some expensive items. A customer who upgrades from a regular bathtub to a whirlpool tub could be increasing the cost of a home by thousands of dollars. If you don't have authorization in writing to make this change, you might never get paid for the increased cost of your plumbing contract.

I can't remember a single house that I've ever built where a customer didn't request some type of change or addition to the original contract. In my early years, I would accept a verbal authorization for changes, but not any more. Too many times I did the work as requested and never got paid. Collecting your money for extras can be difficult enough when you have written documentation of authorization, and it can be almost impossible without some type of proof. A lot of builders lose a lot of money by not putting all their agreements in writing. Don't allow yourself to become another statistic in the book of builders who lost money.

PRO POINTER

If your customers ask for added services or changes in contract obligations, use change orders to document the requested change or additional service you are asked to provide. Otherwise, you could be left holding the bag on some expensive items.A customer who upgrades from a regular bathtub to a whirlpool tub could be increasing the cost of a home by thousands of dollars. If you don't have authorization in writing to make this change, you might never get paid for the increased cost of your plumbing contract.

Be Careful with Partners and Family Members

Be very careful of working with partners and family members. I have worked with partners a few times, and most of the ventures did not have happy endings. Some people do better than others with partners, but I have not known many building partnerships that have lasted more than a few years.

Friends and family members raise a red flag for me when it comes to working together. Not only are you assuming the normal risk of being a builder; you are adding the stress of keeping your friends and family on good terms with you if something goes wrong with the business. Whenever you can swing it, maintain full control over your company.

Pay Your Taxes

Unpaid taxes are responsible for a lot of miserable contractors. It is often tempting to put off paying your taxes until the next good job comes along. This is bad business. The tax debt is not going to go away. This is equally true of payroll taxes and income taxes. Invest in a good Certified Public Accountant and play by the rules when it comes to tax time.

Establish a Relationship with a Good Lawyer

Establish a relationship with a good lawyer when you begin your business. You can cut some corners and save a few bucks by skipping the services of an attorney, but it could be a very poor decision. You should have legal advice in setting up your business. I've known plenty of contractors who have avoided the legal fees in the beginning and ended up with much steeper legal expenses later in their careers.

Be wary of generic forms. Many builders use standard forms. I have seen situations in court where generic forms were deemed to be legal but unenforceable. The end result in these particular cases was a loss for the contractor. Spend the money now to have your lawyer draft suitable documents for your business.

PRO POINTER

Establish a relationship with a good lawyer when you begin your business. You can cut some corners and save a few bucks by skipping the services of an attorney, but it could be a very poor decision. You should have legal advice in setting up your business. I've known plenty of contractors who have avoided the legal fees in the beginning and ended up with much steeper legal expenses later in their careers.

Join Professional Organizations

Join some organizations and participate in them. The growth of sustainable building is huge. With so much to learn, you will benefit from talking with other professionals. I just talked to a builder today who is strong evidence of this. He has built hundreds of homes, but he is just getting into the sustainable side of building. His knowledge has come from research, organizations, and vendors. Mix and mingle to get the most out of current trends.

Avoid Shortcuts

Avoid shortcuts. Most of them do not work out well. It is one thing to be creative and resourceful, but it is very different to cut corners to save time or money. Owning a building business is a big responsibility. Don't take on the job unless you are willing to do it right.

Never Get Too Comfortable

Never get too comfortable with your business. Builders who think they know it all often find out quickly that they don't. If you slacken up, the competition will take advantage of your reduced effort. To enjoy continued success year after year, you can never stop improving your business. Some businesses, such real-estate brokerages, are required to maintain continuing-education requirements. You may not fall under such regulations, but you owe it to yourself to always improve your knowledge and understanding of what it is that you do for a living. If you stop learning, you will likely see a decline in your profits. At the very least, you probably won't see any increase in your business. Regardless of how long you work as a builder, there are always new things to learn.

Matching Green Projects with Suitable Locations

Finding the best building lots is a cornerstone of success for homebuilders. You can have a long line of willing homebuyers standing outside your office and never make a penny if you can't find and secure building lots that suit their needs. You might think that finding lots is no problem, but it can be a big headache. This is especially true if you are working in a rural area with first-time buyers who need cheap land to build on.

My work as a builder has been done in different types of markets. I started out near a big city. Lots were readily available to builders who could commit to buying them in high volume, but spot lots (individual lots in or out of subdivisions) were hard to come by, and they were very expensive.

When I moved my building operation to a less-populated area, lots of all sizes, shapes, and prices were abundant. Finding a lot for any type of buyer was easy. Now that I've moved to Maine, I'm faced with difficulty in finding lots for affordable housing. There is plenty of land in Maine, but most of it is in large parcels, and much of it, large or small, carries a hefty price tag.

I recall one occasion when my advertising was working particularly well. In one week 63 prospects responded to my ads. All the people were looking for starter homes. It was no problem for me to build the houses at prices they could afford, but I could find only two lots that were priced in a range that would work with affordable housing. It pained me greatly to watch 61 disappointed prospects leave my office. Not all the people were qualified to buy a house, and I'm sure many of them were just curious about homeownership, but the lack of lots undoubtedly cost me a bundle of money.

Not All Lots Are Created Equally

Not all building lots are created equally. Some are clearly better than others. In Maine, full basements are very popular among homebuyers. A majority of houses are built on full, buried basements. But not all building lots are suitable for homes with basements. Much of the land in my region is full of bedrock, or what Mainers call ledge. To build a basement on a lot full of ledge requires blasting, and that gets very expensive. An inexperienced builder could walk into a world of trouble if a price was given to a customer for a home with a basement and then ledge was discovered a few feet down in the ground.

Maine has ledge, and Virginia has soggy soil and other building hazards. I'm sure that other parts of the country present their own special types of obstacles for builders. Buying a lot that has hidden expenses associated with it is a sure way to give up your profit on a job. You have to be knowledgeable about what to look for, and you have to be judicious in the lots you select.

Not all hazards are created by nature. Some are related directly to human beings. For example, you might buy a lot that looks great today and find out halfway through the construction of a spec house that an airport is being built nearby and that the flight pattern will be right over your new house. Something like this can make the value of your property plummet.

Before you buy any land, you have to go through a series of checks. Routine real-estate requirements, such as checking to see that the land can be transferred with a clear title, are to be expected. You should check zoning requirements and all other aspects of risk associated with buying real estate. Any good real-estate lawyer can walk

PRO POINTER

Before you buy any land, you have to go through a series of checks. Routine real-estate requirements, such as checking to see that the land can be transferred with a clear title, are to be expected. You should check zoning requirements and all other aspects of risk associated with buying real estate. Any good real-estate lawyer can walk you through the typical concerns. But as a builder, you have to look for more than just the typical legal problems. You must assess the land for its potential to be built on.

you through the typical concerns. But as a builder, you have to look for more than just the typical legal problems. You must assess the land for its potential to be built on.

When you want to build sustainable, environmentally friendly housing, there are other obstacles to overcome. You may remember when solar homes were the talk of the town. The location of a lot is still very important in green building. Imagine the advantages of a good southern exposure when keeping utility costs and usage down. Choosing your lots carefully will result in better success as a green builder.

PRO POINTER

When you want to build sustainable, environmentally friendly housing, there are other obstacles to overcome. You may remember when solar homes were the talk of the town. The location of a lot is still very important in green building. Imagine the advantages of a good southern exposure when keeping utility costs and usage down. Choosing your lots carefully will result in better success as a green builder.

Utility Hookups

One of the first things that a builder must take into consideration is the availability of utility hookups. This is not always a simple issue, but it is always an important one. You might be surprised at how often building lots are affected by utility problems.

Water

Houses require water. If you are building in cities and towns, you probably expect a water main to be within easy reach of a building lot. When an existing water main runs along a property line, it makes life simple for builders. A phone call to the proper authorities will tell you right away how much it is going to cost to get water to the house you are building. But there can be some hidden expenses, and it's up to you to avoid costly surprises.

Under normal conditions with a municipal water main, a branch pipe is run from the main to a location about 5 feet inside the property line of a building lot. The placement of this feed pipe could be in the front or back of the building lot or even on one of the sides. Most water mains are installed under roads, and it is logical to assume that the water hookup will be located in proximity to the road, but this is not always the case.

When a building lot that will be served by a municipal water main is purchased, a tap fee is usually required. The tap fee is not part of the price of the building lot. It is a fee that is required for the privilege of connecting to the city water system. An important fact to remember is that the tap fee doesn't always guarantee that a water connection will be placed on a lot for the price of the fee. Sometimes a contractor will be required to make the actual connection, with the fee only covering the privilege to tie into the main.

Tap fees vary from city to city. While living in Maine for the last eight years I have not built a house that was served by town water. The last tap fees I paid were in Virginia. This was back around the middle of the 1980s. Even then the tap fees were calculated in the thousands of dollars. The price you have to pay to gain access to potable water can be steep. I expect the fees are higher in some locations and lower in others, but it is definitely an expense that you must factor into your job cost. Remember, the tap fee doesn't always mean that a connection will be made and placed on a lot for you. It's very possible that you will have to pay a private contractor to make the actual water tap.

If you are responsible for cutting a road surface and repairing it to gain access to a water main, be prepared for some very expensive costs. Opening a road up is no big deal, but repairing it to town, city, or state standards can cost a small fortune. It's entirely possible for builders to lose most of their profits if they neglect to figure in the cost of getting water to a building lot.

PRO POINTER

When a building lot that will be served by a municipal water main is purchased, a tap fee is usually required. The tap fee is not part of the price of the building lot. It is a fee that is required for the privilege of connecting to the city water system. An important fact to remember is that the tap fee doesn't always guarantee that a water connection will be placed on a lot for the price of the fee. Sometimes a contractor will be required to make the actual connection, with the fee only covering the privilege to tie into the main.

Sewer Connections

Sewer connections in urban areas are much the same as water connections. Tap fees are generally charged for the permission needed to tap into a sewer main. The cost of a sewer connection is likely to be as much the cost of a water tap. Don't overlook this type of soft cost when you are working up a building budget.

Electrical Connections

A modern home without electrical connections isn't worth much. Builders who work in cities tend to take their electrical connections for granted. Rural builders know better than to assume that electrical service is available. They also know that the cost of getting an electrical hookup can be substantial.

There are two types of electrical service that you may have to deal with. You know that regular electrical service will be needed. And you might have a need for temporary power to use while building a home. Getting temporary power can cost several hundred dollars and take weeks to arrange. If you will use generators on your jobs for power, you can skip the temporary setup. Once the house you are building is in dried-in condition, meaning that the interior of the home is protected completely from weather, you can have permanent power installed and use it to complete your construction.

If you are used to working in urban areas, you probably wouldn't give much thought to getting electrical service to a house. Your biggest concern would be whether to have the power lines brought in overhead or buried underground. Builders who work out in the boonies, like I now do, have to make sure power is available and what the cost for getting it to a building lot will be.

When I built my most recent personal home, I was required to sign an agreement with the power company that said I would pay a monthly fee every month for more than two years to cover the cost of having power lines extended to my land. I live about half a mile down a private road, so I'm being charged for the cost of installing poles

PRO POINTER

There are two types of electrical service that you may have to deal with. You know that regular electrical service will be needed. And you might have a need for temporary power to use while building a home. Getting temporary power can cost several hundred dollars and take weeks to arrange. If you will use generators on your jobs for power, you can skip the temporary setup. Once the house you are building is in dried-in condition, meaning that the interior of the home is protected completely from weather, you can have permanent power installed and use it to complete your construction.

and wires to my home. This kind of expense can sneak up on you if you're not thinking about utility costs.

If you or your customers will have to pay thousands of dollars to obtain electrical power, phone service, or cable television, a cheap building lot might not be as good a deal as you think it is.

Wells

Building lots in rural areas are often required to have their own wells as private water sources. The good news is that there is no tap fee to be paid to a city, and there will not be monthly water bills for the rest of a homeowner's life. The bad news is that wells and pump systems are not cheap. A typical well and pump setup can run several thousand dollars. This type of expense has to be factored into the cost of a building lot.

Septic Systems

Septic systems, like wells, are common in rural locations. A simple septic system might cost less than $7,500, but a complex system could cost more than $18,000. When you compare these figures to the cost of a sewer hookup fee, it's easy to see that country lots that look to be inexpensive may just be a nightmare waiting to happen. You have to be diligent in discovering all costs associated with a lot and the construction of a house on it before you can determine if the property is a good value. And never forget that not all land is suitable for a septic system. This can be a real deal-breaker.

Lay of the Land

The lay of the land on a building lot can affect the cost of building a house. A lot that slopes off in any direction will require more foundation work than a level lot will. In some cases, the slope can result in having to spend considerably more for a foundation. Keep this in mind the next time you are walking a building lot and thinking about buying it. At the same time, a sloped lot can be used to your advantage in green building. Digging a portion of a home into a hillside can produce strong benefits for heating and cooling requirements.

Soggy Ground

Soggy ground can give builders problems during and after construction. Land that is soggy is tough to install a foundation on. You must dig down to a depth where solid, dry ground can be found. If you have to go very deep, you've increased the cost of your footings and your foundation walls. After a house is built on this type of land, moisture problems in or under the home can plague your customers, who will in turn hound you. Be very careful if you are buying soft ground.

Rock

Buying a building lot that is hiding rock just below grade can cause you all sorts of problems if you are required to install a septic system or a basement. Blasting bedrock is an expensive proposition. If you work in an area where rock is likely to be a problem, do some probing in the ground with a steel drive rod or have a back-hoe operator dig a few test pits for you. Don't agree to

> ***PRO POINTER***
>
> Buying a building lot that is hiding rock just below grade can cause you all sorts of problems if you are required to install a septic system or a basement. Blasting bedrock is an expensive proposition. If you work in an area where rock is likely to be a problem, do some probing in the ground with a steel drive rod or have a back-hoe operator dig a few test pits for you. Don't agree to build a basement or install a septic system before you know whether rock is going to stop you.

build a basement or install a septic system before you know whether rock is going to stop you. Even putting in a normal sewer and water service can be very difficult if rock is encountered. Look before you leap. And, keep in mind that bedrock can mean that there could be problems with radon when you build a home.

Flood Zones

Buying building lots that are located within established flood zones is very risky. The land might be buildable, but financing can be very difficult to arrange for properties located in flood planes and flood zones. If a property is situated in a flood hazard area and financing is needed, flood insurance will almost certainly be required by the lender. This type of insurance is expensive. There is also the concern that buyers will not be interested in owning a home where a flood might wipe out everything they own. I'd stay away from flood risks.

Trees

Trees on a piece of land increase the cost of building a home. This cost is usually offset by the desirability of a lot with trees, but you should still be aware that extra costs will be incurred. Having trees cleared can tack hundreds of dollars onto the price of a house. I've known builders who looked for lots without trees to save this expense. Personally, I usually buy wooded lots and pay to have them cleared, because I feel that a wooded setting helps to sell most houses faster.

Access

Access to a building lot might not seem like something that you would have to worry about, but it can be. Suppose you bought a lot that came with a 25-foot driveway easement for access; would it be a good deal at a low price? I doubt it. Most lenders require a property that is accessed by a right-of-way to have an easement that is at least 50 feet wide. Buying a lot with a narrow right-of-way might mean that you or your customers will not be able to obtain conventional financing for construction. Technically, the lot is not inaccessible, but it may not be able to be financed.

Maintenance Agreements

Road-maintenance agreements are not uncommon in many areas. If several building lots are situated along a private road, it is customary for the land developer to create a road-maintenance agreement. Basically, these agreements say that each property owner is responsible for an equal portion of road repairs and maintenance. This may not sound like much when you are looking at land, but the cost of keeping up a road can be steep. Just having my road plowed when it snows costs me $100 a storm. Adding layers of stone is very expensive, and the cost of paving can be astronomical. For this reason, some road-maintenance agreements can be red flags that you should run away and look for a lot in some other location.

PRO POINTER

Road-maintenance agreements are not uncommon in many areas. If several building lots are situated along a private road, it is customary for the land developer to create a road-maintenance agreement. Basically, these agreements say that each property owner is responsible for an equal portion of road repairs and maintenance. This may not sound like much when you are looking at land, but the cost of keeping up a road can be steep.

Dues and Assessments

Some housing developments have a structured plan in which homeowners must pay dues and assessments for maintenance and amenities. This is not at all uncommon in planned developments. The homeowner dues can go up year after year. Building in subdivisions where homeowner dues are a requirement might hurt your chances of selling a spec house. There are certainly a lot of people who are willing to assume financial responsibilities in return for good roads, playgrounds, parks, swimming pools, and similar amenities, but you should at least be aware that the costs of a property-owner association could lose you some sales.

Restrictions

Many subdivisions contain lots that are controlled by covenants and restrictions. This is something that your lawyer should look for when assessing the deed and title of a piece of land. As a builder, you must ask what the restrictions are. For example, many subdivisions require any house built to contain a minimum amount of living space. It could be 1,500 square feet, 1,800 square feet, or more. If you are planning to build a ranch-style home, you might learn that the subdivision will not allow any one-story homes. The restrictions can be very detailed.

PRO POINTER

Many subdivisions contain lots that are controlled by covenants and restrictions. This is something that your lawyer should look for when assessing the deed and title of a piece of land. As a builder, you must ask what the restrictions are. For example, many subdivisions require any house built to contain a minimum amount of living space.

I've built in subdivisions where roof colors and types were regulated. So were the color and type of siding that could be used. Even the foundation materials are sometimes specified in the restrictions. This is done to maintain a quality subdivision, but it can be a shock to an unsuspecting builder. Make sure that there are no prohibitions on the type of house you plan to build before you agree to buy a building lot.

Finding the Cream of the Crop

When you're looking for building lots, finding the cream of the crop can take a lot of work. How you go about locating lots will depend somewhat on the type of builder you are. High-volume builders usually make deals with land developers to acquire control over many lots on a graduated takedown schedule. Some builders buy raw land and develop their own lots. Done properly, this can be a very profitable venture. However, it can also be the downfall of a good construction company.

If you are going to build only a few homes a year, you will most likely find your lots in a more traditional manner. Advertisements in your local newspaper are a good place to start. Talking to real-estate professionals can lead to some good lot acquisitions.

Buying lots from ads in the paper or from signs that you see in various locations is a common way of doing business as an average builder. But it is not always the best way to find and secure ideal building lots. Sometimes you have to dig deep and turn up lots that are not advertised for sale. In fact, many of the lots you go after may not even be for sale when you start making offers to buy them. Let's expand on this a bit.

Unadvertised Specials

Some builders are excellent at finding unadvertised specials. If you learn to become one of these builders, you will enjoy better lots, better prices, and less competition. When a building lot is advertised in a newspaper or placed in a multiple-listing service, the demand for the lot can be great. Sellers who are seeing a lot of activity may hold out for top dollar, and they may not be willing to negotiate special purchase or option terms. If you can locate suitable lots that are either not for sale but might be able to be bought or lots that are getting very little exposure on the open market, you're in a better buying posture.

Developers of subdivisions frequently have some lots that simply don't sell during the major sales effort of a subdivision. These lots become leftovers, and they can be an excellent value for a small-volume builder. I have built houses on numerous leftover lots, and with the exception of one lot, I've always done well with the ventures. The one lot that I didn't do so well with had underground water problems that I didn't anticipate, which ran up the cost of construction and ate into my profits.

Finding leftover lots is easy. Get in your truck and ride around to look at some established subdivisions. Take note of any lots that have not been built on. Determine their address or legal descrip-

PRO POINTER

Finding leftover lots is easy. Get in your truck and ride around to look at some established subdivisions. Take note of any lots that have not been built on. Determine their address or legal description and make a trip to city hall to find out who the owners are. If there are lots sitting empty in a mature subdivision, there's a very good chance that you can buy them at a good price, especially if they are still owned by the developer.

tion and make a trip to city hall to find out who the owners are. If there are lots sitting empty in a mature subdivision, there's a very good chance that you can buy them at a good price, especially if they are still owned by the developer.

By checking public records in the local tax office or courthouse, you can see who owns what and what they paid for it when they purchased it. A little more detective work will reveal an address and maybe a phone number. Once you know how to make contact, you simply call or write the owners and let them know of your interest in buying their land. This gets the ball rolling, and

PRO POINTER

Many times people will sell their land if asked to, but they have not yet gone to the trouble to find buyers. Developers sometimes lose interest in trying to sell lots that just don't seem to sell, so they stop advertising and move onto new projects. In either case, sellers might be willing to do business with you on discount terms.

the worst that can happen is that you will receive no response or be told that the land is not for sale. In the best case, you will get a quality lot at a bargain price.

Many times people will sell their land if asked to, but they have not yet gone to the trouble to find buyers. Developers sometimes lose interest in trying to sell lots that just don't seem to sell, so they stop advertising and move onto new projects. In either case, sellers might be willing to do business with you on discount terms. And speaking of terms, you might be able to arrange some attractive purchase options, since the sellers are not in a big hurry to make a sale. This can help your cash flow and still give you control over lots that can be built on.

Small Publications

Small publications often allow people to place free or nearly free ads to sell their possessions. This type of publication has existed in every state that I've lived in. Some real-estate brokers advertise in these little papers and booklets, but most brokers spend their time and money with more aggressive markets that have larger circulation rates. The deals in these little buy-and-sell papers are frequently average at best, but there are times when a really wonderful opportunity shows up. I bought the land for my personal home from an ad in such a publication, and I got a great deal on it.

Many builders look in their local papers and deal with real-estate brokers who are members of multiple-listing services. Some builders dig deeper, but many don't. By watching obscure publications, you just might stumble across some excellent opportunities. Don't expect to find them fast. It may take months. But, if you are diligent in your search, something worth buying will show up eventually. It always has for me.

Brokers

If you develop a quality relationship with a few outstanding brokers, you might enjoy some fabulous land deals. Many builders shun brokers for various reasons. I'm both a builder and a broker, so I know the game from both sides of the table. There are a lot of inadequate brokers and agents offering their services in the real-estate industry. Don't judge all real-estate professionals by the actions of a few. The right brokers can work wonders for you both in acquiring first-class lots and in making sales for your new homes.

Do It Yourself

We've all heard the old saying "If you want something done right, you have to do it yourself." There's a lot of truth to this statement. If you're capable of doing a job yourself, it's difficult to find someone whom you feel can do the work better. This applies to painting, carpentry work, and land development. Becoming your own developer is a big and often risky decision, but it can pay big dividends. As an alternative, you might do well to work with some land developers to obtain exclusive rights to good lots.

Controlling Desirable Lots without Buying Them

Controlling desirable lots without buying them is a great way to operate. Doing this is easier than you might think, but it is not without risk. You must approach this method of lot acquisition with the same respect that you would when buying lots straight out.

As a volume builder, I've often worked out deals with developers so that my company controlled dozens of building lots with very little actual cash being tied up. When I was developing subdivisions on my own, I worked out deals with other builders to allow them an opportunity to control lots without buying them. I've made the deals as both a developer and as a builder, so you have the advantage of hearing how options and takedowns work from both perspectives.

When land developers create subdivisions, they need builders to come in and buy the lots that have been created. Few developers depend on the general public to buy their lots, although some will sell to anyone who has enough money. In most cases, developers contact established builders and try to sell them all the available lots. It is not expected that one builder will take over an entire subdivision, but this does happen from time to time. Usually, three or four builders are solicited to build out a subdivision. Wise developers pick their builders carefully.

A developer likes to see builders working together to create a harmonious housing development. Green builders will be sought by green developers. Wise developers look for builders who will complement each other without providing direct competition. Some developers have little concern over who builds what as long as the lots sell. There is no tried and true rule of how the sellout of a subdivision works. But, it is a reasonably consistent fact that developers will seek out builders to buy a majority of all lots created. This is where you come in.

PRO POINTER

When land developers create subdivisions, they need builders to come in and buy the lots that have been created. Few developers depend on the general public to buy their lots, although some will sell to anyone who has enough money. In most cases, developers contact established builders and try to sell them all the available lots. It is not expected that one builder will take over an entire subdivision, but this does happen from time to time. Usually, three or four builders are solicited to build out a subdivision. Wise developers pick their builders carefully.

You can wait for a developer to contact you, or you can go directly to the developers and solicit them. If you don't have a track record as a builder, some developers may be apprehensive about doing business with you. Developers want builders who can uphold their takedown schedules and build out a subdivision in a timely fashion.

Takedown Schedules

If you're not familiar with takedown schedules, I will explain them to you. They're really quite simple. Let's say that I'm a developer and you are a builder. As part of my routine, I normally won't sell lots in less than a 20-lot package. My building lots sell for $60,000 each. To buy all the lots at one time, you would need $1,200,000. Few builders can handle this type of up-front expense, and there are not many lenders who will take such a big gamble on a new subdivision. This leaves me, the developer, with a problem. I've got 80 lots to sell, and I can't find builders who are flush enough to buy 20 at a time. What am I going to do?

In this situation, I'm going to offer you a takedown schedule. You will sign a contract that says you will purchase 20 lots for $50,000 each. I will get an earnest-money deposit when the contract is signed. The amount of the deposit might be $500 or less, but it could be much more. It's whatever the two of us agree to. Then you buy your first lot right away. That leaves 19 lots to be purchased over some period of time. To determine how many lots you must buy in a given period of time, we create a takedown schedule. You might buy one lot each month, or you might buy a new lot every time the lot you are

PRO POINTER

Takedown schedules are created frequently between builders and developers. They are about the only practical way for developers to make large sales consistently. Builders like using takedown schedules because they allow them to offer their customers 20 lots, or whatever the number, to choose from, even though they may actually only own one of the lots. It's a win-win situation for both the builder and the developer.

building on at the present time is sold. The takedown schedule might call for you to buy all the lots within a 12-month period regardless of when you buy which lots. The options for terms in a takedown schedule are limited only by the agreements made between you and me.

Working with takedown schedules is good for both of us. You get control of 20 lots with a minimal amount of out-of-pocket cash. I get a commitment for 20 sales that I can show my banker. This frees me up to keep working with a new line of credit. I don't have to peddle my lots one at a time to the general public. If you default, I keep the deposit money, maintain ownership of the unpurchased lots, and sue you for your lack of performance. Assuming that the deal works out well, we both win. You have control of exclusive lots that you can buy gradually, and I have sold a chunk of my subdivision.

Takedown schedules are created frequently between builders and developers. They are about the only practical way for developers to make large sales consistently. Builders like using takedown schedules because they allow them to offer their customers 20 lots, or whatever the number, to choose from, even though they may actually only own one of the lots. It's a win-win situation for both the builder and the developer.

As I said earlier, there is risk to takedown schedules. Since little money changes hands when a deal is struck, it's easy for builders to get greedy. They may sign up for more lots than they can reasonably buy in a set period of time. If a builder defaults, a developer is likely to attack with legal proceedings. This is bad news for the builder, both from a public-image point of view and from a financial perspective.

You could go into a new subdivision with a great deal of optimism and secure 20 lots on a takedown schedule, only to find that the public is not enthralled with the housing development. Depending upon the terms of your takedown schedule, you might have to buy a lot each month even though you are unable to sell the houses needed to fill them. This can be a crushing blow. There is a thin line between not committing to enough lots and making a commitment that will sink your building business. Knowing what to do is part luck and part experience. If you have enough experience and invest enough time in researching local demographics, you can create your own luck.

Options

Land options are another way to control property without paying for it. Once you have contractual control over a lot, you have an advantage over other builders. If a customer falls in love with a lot that you control, it's impossible for any other builder to build that customer's home unless you are willing to settle for just making a lot sale. The more lots you control, the less competition you have.

Options, like takedown schedules, are not governed by set rules. A buyer and seller can work out any terms that the two of them are comfortable with when setting up a purchase option.

PRO POINTER

Land options are another way to control property without paying for it. Once you have contractual control over a lot, you have an advantage over other builders. If a customer falls in love with a lot that you control, it's impossible for any other builder to build that customer's home unless you are willing to settle for just making a lot sale. The more lots you control, the less competition you have.

Investors and builders often use options to gain control of real estate with minimal cash demands. If you smell a great opportunity but aren't completely sure of yourself, an option is a good choice. The most you stand to lose when you option the purchase of a piece of property is the money that you put up for the option. Let me explain how some types of options might work and why they're advantageous to you.

Short-Term Options

I describe short-term options as options that must be acted on within 90 days. Three months is not a lot of time, but it can be enough for a builder to make evaluations that could make or break a big deal. Setting up a short-term option is easier than working out the details of some longer options. If the seller of a piece of property is anxious to sell, it's unlikely that any option will be considered too strongly. Once an option is placed on a piece of property, the seller's hands are tied until the person holding the option decides to act. But short-term options can often be placed with developers and investors who are accustomed to playing the real-estate game. Typical consumers who are selling their personal land are less receptive to option angles.

When you offer a seller terms for an option, you can make the offer in any form that you wish. The seller isn't obligated to accept your offer and may counter it with a different set of terms. It never hurts to ask for an option. The worst that can happen is that you won't get it. Since you are more likely to be interested in long-term options as a builder, let's discuss the details of all basic option plans with the longer term in mind.

Basic Options

Basic options for builders usually run for longer than 90 days. A six-month option is not unusual. Four months is very common, and other terms exist. To illustrate how this type of deal is done, let's assume our roles with you as a builder and me as a developer. You want to control three lots in my subdivision, but you're not comfortable with committing to a purchase contract and a takedown schedule until you can test the public opinion of my new subdivision.

As a developer, I plan to sell all my building lots within a three-year time period. You contact me and offer to option three of my lots for six months. In the offer, you agree to give me $5,000 in nonrefundable option money. If you don't buy the lots, I keep the five grand. If you do buy them, the money is applied to the purchase price. I would prefer a contract to buy and a takedown schedule, but I have a lot of property to sell and I have three years to sell it in. Taking your money now might result in three sales. At the worst, I'll get $5,000 and still have my lots to sell later. Under the circumstances, I counter your offer with an offer that requires you to put up $10,000.

PRO POINTER

Basic options for builders usually run for longer than 90 days. A six-month option is not unusual. Four months is very common, and other terms exist.

You now have an opportunity to control many valuable lots for only $10,000. It's your call. If you take the option, advertise the lots with houses built on them, and achieve quick sales, you can option more lots and maintain your momentum. You decide to do the deal. Options are as simple as this. The most you stand to lose is your $10,000. You're not getting in over your head by committing to a major deal, but you are gaining contractual control over three very desirable lots. It's not a bad deal. But you decide to make the deal a little sweeter for yourself before you let me know that you would take the deal as it stands if you have to.

You counter that you will put up the $10,000 but that for every lot purchased, one-third of the money is applied to the purchase price. This way, if you buy and sell one lot, the most you can lose is a portion of your option money. If you buy two lots, you're at half the risk. Since I have plenty of lots and time, I accept your offer and we cut the deal. You're off and running with very little cash out of pocket and minimal risk. This is the way that smart builders test the water before they jump in with both feet.

> **PRO POINTER**
>
> Basic options allow both buyers and sellers to arrive at terms that suit each other. There are no real rules to the game. Offers go back and forth until a deal can be struck. If you can find sellers who are not panicked for quick cash, options are very effective. When making offers on leftover lots and lots that are not officially for sale, sellers have very little to lose by accepting options. Keep this in mind the next time you are shopping for lots. If your gut instincts turn out to be wrong, you will have lost your option money but not your shirt.

Basic options allow both buyers and sellers to arrive at terms that suit each other. There are no real rules to the game. Offers go back and forth until a deal can be struck. If you can find sellers who are not panicked for quick cash, options are very effective. When making offers on leftover lots and lots that are not officially for sale, sellers have very little to lose by accepting options. Keep this in mind the next time you are shopping for lots. If your gut instincts turn out to be wrong, you will have lost your option money but not your shirt.

Stretch Your Money

Learning how to stretch your money with options and takedown schedules is very important for spec builders. Tying your money up in lots will reduce your ability to handle the carrying cost of keeping houses on the market for longer than the period of time that you had budgeted. Even if you are building custom homes, it can be quite beneficial to have a diverse stable of building lots available only to you.

When you are first starting out, buying and optioning lots might be too risky. You can advertise your services and hope to find customers who already own their own lots. Some customers do. You can also assume that the customers you take on in the early stages of your business will find suitable lots on the open market. I can't tell you that you will or won't be successful by locking up building lots. It is a fact that the person who controls the land controls the building, so keep this in mind. Weigh your opportunities carefully and don't overextend yourself. Most builders who fail do so because they leverage too much on credit. Building your business slowly may be agonizing, but it is often the best route to take. You have been given a lot of solid information, and you will get a lot more in future chapters, but you have to mold your new knowledge to fit your personal circumstances. What works for me might not work for you.

Finding Financing the Easy Way

Finding financing is essential for most builders. You can do it the hard way or the easy way. There is no truly easy way, but there are methods that require less effort and work better than others. These are the ones I have always preferred. There is a catch. If you don't have the experience or the contacts, the easy path to financing is not so easy.

Establishing a line of credit is a task most business owners must tackle. Setting up credit accounts for a new business can be tough for anyone, but for people with poor credit histories, the job can seem almost impossible. Vendors don't want to establish credit unless you have a good track record, but how can you develop a good track record if people won't give you a chance? This is kind of like the chicken and the egg—which one comes first? This chapter is going to help you to understand the methods used to establish credit.

Good Credit Is Crucial to a Growing Business

Good credit is crucial to a growing business. Without good credit, expansion, even survival, can be very difficult. Most businesses need credit accounts, whether credit is used on a monthly basis to buy supplies or on a semiannual basis as operating capital. If you have an existing business without any credit blemishes, setting up new accounts is not too difficult. If, on the other hand, you have a poor credit rating, getting new credit lines can be an uphill battle. For a new business, establishing credit can be a slow process, but it can be done.

What types of credit will you need? As your business grows, so will your desire for various types of credit. Let's look at some of the types of credit you may have a need for.

Start-Up Money

The first need you may have for credit could be to obtain start-up money. Start-up money is the money you will use to get your business started.

It is best to rely on money you have saved for start-up capital. Borrowing money to start your business will put you in a hole right from the start. The burden of repaying a loan used for start-up money will only make establishing your business more difficult. However, many successful businesses are started with borrowed money.

When I started my first business many years ago, I had to borrow $500 just to buy tools to fill my toolbox. Today $500 doesn't seem like much money, but back then it felt like a big risk to take. But I took the risk and it paid off. While I can't recommend going into debt to start your business, I did it and it worked.

A loan for start-up capital can be one of the most difficult to obtain. Lenders know that many new businesses don't survive their first year of operation. If you don't have home equity or some other type of acceptable collateral, banks will be reluctant to help you get your business started.

If you wait until you have quit your job to apply for a start-up loan, chances of having the loan approved drop considerably. Most savvy entrepreneurs arrange a personal loan while they have their regular job and use these funds for their business start-up. This procedure results in more loan approvals.

Some contractors start their businesses on a part-time basis. By doing this, they are able to obtain the loans they need while they are still employed and can build up cash reserves. The work they do on the side produces money to repay their loans and can accumulate into a healthy stockpile of

PRO POINTER

It is best to rely on money you have saved for start-up capital. Borrowing money to start your business will put you in a hole right from the start. The burden of repaying a loan used for start-up money will only make establishing your business more difficult. However, many successful businesses are started with borrowed money.

ready cash. The stress and hours of working full-time and running a part-time business can be tiresome, but the results are often worth the struggle.

Operating Capital

Once your business is open, you will need some operating capital—money used to bridge the gap between job payments and to keep the business running. If you don't already have a supply of money set aside for this purpose, you may want to apply for an operating-capital loan.

Lenders are a little more willing to make these loans if you have an established track record. However, don't get your hopes up that you will be approved for any loan as a self-employed person until you can provide two year's tax

returns for income verification. Since you will probably need operating capital to survive your first two years, make arrangements for your financing before you jump into business.

Operating-capital loans can be provided from several types of financing. Some business owners borrow against their home equity to generate operating capital. Lines of credit are often set up for the contractor to use money as needed. With this type of financing, you only pay interest on the money you are using. Short-term personal loans are another way of financing your operating capital. These loans are frequently set up with interest-only payments until the note matures, at which time the total amount becomes due.

Trade Accounts

You will most likely need a variety of trade accounts. These are financing arrangements that make doing business much easier. They are used for everything from advertising to products. The accounts buy you a little time. However, they can sink your financial ship quickly. Running up high trade accounts without paying them off as income is collected is a recipe for trouble.

Advertising Accounts

Advertising credit accounts are convenient. These accounts allow you to charge your advertising and pay for it monthly, thereby eliminating the need for cutting a check for every ad you place. It also gives your advertising a chance to generate income to pay for itself. However, be careful; if you abuse these accounts, you can get into deep debt quickly.

Although advertising often pays for itself, sometimes it doesn't. You cannot afford to charge thousands of dollars' worth of advertising that doesn't bring in paying customers. If you do, your business will be hurt before it starts by having to pay for ads that are not effective. Use your credit accounts prudently.

Supplier Accounts

Supplier accounts are convenient. By establishing credit with material suppliers, you will not have to take your checkbook with you every time you need supplies.

Since you will probably have to supply and install materials before your customers will pay you, supplier accounts buy you some float time, generally up to 30 days. You will be able to get the materials you need without paying for them until the first of the next month. This gives you time to install the materials and collect from your customers before paying for them.

Again, you must be cautious. If your customers don't pay their bills, you won't be able to pay yours. You must keep the reins tight on collecting your accounts receivable. Once you have opened an account and used it, you must be prompt in paying your bill when it comes due in order to continue the process of developing credit.

PRO POINTER

Since you will probably have to supply and install materials before your customers will pay you, supplier accounts buy you some float time, generally up to 30 days. You will be able to get the materials you need without paying for them until the first of the next month. This gives you time to install the materials and collect from your customers before paying for them.

Office Supplies

You may be able to get better pricing on your office supplies by purchasing them from mail-order distributors. Having a credit account with these companies will make your life easier by avoiding the need to pay COD for each delivery. With a business account, you won't have to put business expenses on your personal credit cards. This type of account makes your accounting easier and your business less troublesome to keep up with.

With discount office-supply companies like Staples or Office Depot, you will find it easy to open an account for all your office supplies, and by paying promptly when the bills come due, you'll be on your way to establishing good credit.

PRO POINTER

With discount office-supply companies like Staples or Office Depot, you will find it easy to open an account for all your office supplies, and by paying promptly when the bills come due, you'll be on your way to establishing good credit.

Fuel

Paying cash for fuel for your trucks and equipment can be a problem when you have employees who need gas or oil when you are not around. Keeping up with the cash and the cash receipts is time-consuming. By establishing a credit account with your fuel provider, your employees can charge fuel for your company vehicles and equipment. At the end of the month, your account statement will be easy to transfer into your bookkeeping system. But you will need a system to avoid abuse of this charge account and make sure that all purchases are for business and not personal use.

Vehicles and Equipment

As your business grows, you may need more vehicles and equipment. Most business owners can't afford to pay cash for these large expenses. Financing or leasing vehicles and equipment will require a decent credit rating.

There are times when credit may make a difference between the survival and failure of your business. Develop a good credit rating and work hard to maintain it.

Selecting Your Lending Institutions

One of the first steps in establishing credit is to select your lending institutions. Not all lenders are alike. Some lenders prefer to make home-mortgage loans. Others make car loans and other secured loans. Some lenders will provide any type of loan if they feel the loan is safe. Finding a bank that is willing to make an unsecured signature loan can be troublesome.

Before you can set out to find a lender, you must decide which type of loan you will be requesting—a secured loan or a nonsecured loan. If you do not have a well-established and financially sound business, be prepared to sign the loan personally. Most lenders will not make a business loan without a personal endorsement. A secured loan is a loan where something of value, collateral, is pledged to the lender for security against nonpayment of the loan. An unsecured loan is a loan that is secured by a signature but not by a specific piece of collateral. Most lenders much prefer a secured loan.

PRO POINTER

A secured loan is a loan where something of value, collateral, is pledged to the lender for security against nonpayment of the loan. An unsecured loan is a loan that is secured by a signature but not by a specific piece of collateral. Most lenders much prefer a secured loan.

If you plan to request a secured loan, you must decide what you have to use as collateral. The amount of money you wish to borrow will have a bearing on the type and amount of collateral that is acceptable to the lender. For example, if you want to borrow $50,000 for operating capital, putting up the title to a $10,000 truck that has been fully paid for will probably not be sufficient.

When you are looking to borrow large sums of money, real estate is the type of collateral most desired by lenders. If you have equity in your home and are willing to risk it, you should find that getting a loan is relatively easy. However, home-equity loans can be confusing. Many people don't understand how much they can borrow against their equity. Let me show you two examples of how you might rate the loan value of the equity in your home.

Let's say your home is worth $100,000 and you owe $80,000 on the mortgage. I know the value given here is low, but the math is easy to do and to understand. Equity is the difference between the home's appraised value and the amount still owed on it. In this case the equity amount is $20,000. Does this mean you can go to a bank and borrow $20,000 against the equity in your home? No, not likely; most lenders would consider such a loan a high risk. Some finance companies might lend $10,000, but they would be few and far between. In most cases you would not be able to borrow any money against your equity.

Lenders want borrowers to have a strong interest in repaying loans. If a lender loaned you $20,000 on your equity, your house would be 100 percent financed. Basically, if you defaulted on the loan, you wouldn't lose any money, only your credit rating. To avoid this type of problem, lenders require that you maintain an equity level in your house that is above and beyond the combination of your first mortgage and the home-equity loan. Most banks will want you to have between 20 and 30 percent of the home's value remaining in equity even after getting a home-equity loan.

Let's say you have the same $100,000 house, but you only owe $50,000 on it. A conservative lender would allow you to borrow $20,000 in a home-equity loan. This brings your combined loan balance up to $70,000, but you still have $30,000 of equity in the home. If you default on this loan, you lose not only your credit rating but $30,000 as well. A liberal lender may allow you to borrow $30,000, keeping an equity position of $20,000.

If you don't have real estate for collateral, you can use personal property. Personal property might include vehicles, equipment, accounts receivable, or certificates of deposit. Different lenders will have different policies, so shop around until you find a lender you like doing business with.

Banks are an obvious choice when applying for a loan. Commercial banks often make all types of loans. But savings-and-loan companies also deal in real-estate loans and are worth investigating. If you belong to a credit union, check its loan policies and rates. Finance companies are usually aggressive lenders, but their interest rates may be high. Private investors are always looking for viable projects to invest in or to loan money to. Mortgage brokers are yet another possibility for your loan. A quick look through the phone directory and the ads in the local newspaper will reveal many potential loan sources.

> **PRO POINTER**
>
> If you don't have real estate for collateral, you can use personal property. Personal property might include vehicles, equipment, accounts receivable, or certificates of deposit. Different lenders will have different policies, so shop around until you find a lender you like doing business with.

How to Establish Credit When You Have None

This section is going to teach you how to establish credit when you have none. It is better to start with no credit than with bad credit. If you have never used credit cards or accounts, you will find it difficult to get even the smallest of loans from some lenders. Another problem you will encounter is the fact that you are self-employed. Most lenders don't want to loan money to self-employed individuals until the individuals can produce at least two years' tax returns. With this in mind, you might be wise to set up your accounts and credit lines before quitting your full-time job.

Since most people don't set up business accounts until they are in business, I will give you advice on how to get credit without having a regular job. However, if you have the opportunity to establish your business accounts before quitting your job, do it.

Supplier Accounts

Supplier accounts are one of the best places to begin establishing your new credit. When you go into business, suppliers will want to supply you with needed materials. Suppliers will be cautious about granting you a credit account, but they will be more liberal than the average bank.

As a new business you need credit, materials, and customers. As ongoing and competitive businesses, suppliers need new customers. In effect, you need each other; this is your edge. If you handle yourself professionally, you have a good chance of opening a small charge account. Now let's look at how you should go about opening your new supplier accounts.

You can request credit applications by mail or you can go into the stores and pick them up from the credit department. Most applications will want personal references, credit references, your name, address, phone number, Social Security number, bank balances and account numbers, business name, and much more. The application will ask how much credit you are applying for. Don't write that you are applying for a very high limit to start with; pick a realistic figure that is a little higher than what you really want. Setting the higher figure will give you some negotiating room.

Most supplier credit applications are similar. Once you fill out the first application, make copies of it for future reference. Not only will the photocopy serve to refresh your memory if the supplier has questions on your application; it will act as a template in filling out other credit applications.

When you have completed the applications, return them to the credit department. You might want to hand-deliver them; this will give you a chance to make a personal impression on the

PRO POINTER

Most supplier credit applications are similar. Once you fill out the first application, make copies of it for future reference. Not only will the photocopy serve to refresh your memory if the supplier has questions on your application; it will act as a template in filling out other credit applications.

credit officers. Making a good impression on the credit manager will help.

If you get a notice rejecting your credit request, don't give up. Call the credit manager and arrange a personal interview. Meet with the manager and negotiate for an open account. When all else is failing, ask for a smaller credit line. Almost any supplier will give you credit for $500. This may not sound like much, but it's a start, and a start is what you need.

Take whatever credit you can get. After you establish the account, use it. Having an open account is not enough. You must use the credit to gain a good credit rating. You should use the credit account frequently and pay your bills promptly. When the supplier offers a discount for early payments, pay your bills early. By paying early and taking advantage of the discounts, you will save money and improve your credit rating.

Suppliers offer the easiest access to opening new accounts and starting to build a good credit rating. After these accounts are used for a few months, your credit rating will be growing. By keeping active and current accounts with suppliers you are building a good background for bank financing.

> **PRO POINTER**
>
> Suppliers offer the easiest access to opening new accounts and starting to build a good credit rating. After these accounts are used for a few months, your credit rating will be growing. By keeping active and current accounts with suppliers you are building a good background for bank financing.

Other Vendor Accounts

Other vendor accounts are also valuable. When you open your business, you will need office supplies. You will need the services of a printer. Newspaper advertising is normally a part of every new business. All these miscellaneous vendors offer the opportunity for creating credit. It can be easy to set up accounts with small businesses. If you are a local resident of the community, small businesses may not even ask you to fill out credit applications. These opportunities are too good to overlook. But remember, paying promptly is the key to establishing good credit with vendors.

Major Lenders

Major lenders are a little more difficult to deal with. Unless you have a strong credit rating, tangible assets, and a solid business plan, many banks will not be interested in loaning substantial sums of money. But don't despair. There are ways to work with banks.

Banks, like suppliers, should be willing to make a small loan to you. I know $500 will not buy much in today's business environment, but it is a worthwhile start to building a credit rating. There is another way to build your credit rating and get noticed by your banker.

Bankers like to have collateral for loans. What better collateral could you give a banker than cash? I know, you are thinking that if you had cash for collateral, you wouldn't need a loan. But that is not always true. When you are establishing credit, any good credit is an advantage. Let me tell you how to get a guaranteed loan.

Set up an appointment to talk with an officer of your bank. Tell the banker that you want to make a cash deposit in the form of a certificate of deposit (CD) but that after you have the CD on deposit, you want to borrow against it. Many lenders will allow you to borrow up to 90 percent of the value of your CD. For example, if you put $1,000 in a CD, you should be able to borrow about $900 against it. You are essentially borrowing your own money and paying the bank interest for the privilege. This concept may sound crazy, but it will work to build your credit rating, and don't forget your CD will be earning interest all the while.

Banks report the activity on their loans to credit bureaus. Even though you are borrowing your own money, the credit-reporting agency will show the loan as an active, secured loan. As long as you make the payments on time, you will get a good credit rating. This technique is often used by people repairing damaged credit, but it will work for anyone.

Building a good credit rating can take time. The sooner you start the process, the quicker you will enjoy the benefits of a solid credit history. The road to building a good credit rating can be rocky and tiresome, but it's worthwhile.

PRO POINTER

Banks, like suppliers, should be willing to make a small loan to you. I know $500 will not buy much in today's business environment, but it is a worthwhile start to building a credit rating. There is another way to build your credit rating and get noticed by your banker.

PRO POINTER

Building a good credit rating can take time. The sooner you start the process, the quicker you will enjoy the benefits of a solid credit history. The road to building a good credit rating can be rocky and tiresome, but it's worthwhile.

Checklist of loan application needs

- ❑ Home address for the last five years
- ❑ Divorce agreements
- ❑ Child support agreements
- ❑ Social security numbers
- ❑ Two years of tax returns, if self-employed
- ❑ Paycheck stubs, if available
- ❑ Employee's tax statements (i.e., W-2, W-4)
- ❑ Gross income amount of household
- ❑ All bank account numbers, balances, names, and addresses
- ❑ All credit card numbers, balances, and monthly payments
- ❑ Employment history for last four years
- ❑ Information on all stocks or bonds owned
- ❑ Life insurance face amount and cash value
- ❑ Details of all real estate owned
- ❑ Rental income and expenses of investment property owned
- ❑ List of credit references with account numbers
- ❑ Financial statement of net worth
- ❑ Checkbook for loan application fees

FIGURE 5-1   A checklist of loan application requirements.

How to Overcome a Poor Credit Rating

If you are starting with a bad credit rating, this section will give you options on how to overcome it. There is no question that setting up credit accounts will be harder if you have a poor credit history, but you can do it. If you are battling a bad credit report, plan on spending some time in cleaning up the existing report and building new credit. This journey will not be easy, pleasurable, or quick, but the results should make you happy.

Secured Credit Cards

Secured credit cards are one way for people with damaged credit to begin the rebuilding process. Secured credit cards are similar to the procedure described for CD loans. A person deposits a specified sum of money with a bank or credit-card company, and a credit card is issued to the depositor with a credit limit equal to the amount of the cash deposit or slightly more. You are basically borrowing your own money but also rebuilding your credit.

PRO POINTER

Secured credit cards are one way for people with damaged credit to begin the rebuilding process. Secured credit cards are similar to the procedure described for CD loans. A person deposits a specified sum of money with a bank or credit-card company, and a credit card is issued to the depositor with a credit limit equal to the amount of the cash deposit or slightly more. You are basically borrowing your own money but also rebuilding your credit.

CD Loans

If you need to rebuild your credit, CD loans are a good way to do it. By depositing and borrowing your own money, you are able to build a good credit rating without risk.

Erroneous Credit Reports

Erroneous reports on your credit history are not impossible. If you are turned down for credit, you are entitled to a copy of the credit-report information used in making the decision to deny your credit request. If you are denied credit, you should immediately request a copy of your credit report. Credit-reporting bureaus make mistakes. Let me tell you a quick story about how my credit report was mangled when I applied for my first house loan.

When I applied for my first house loan, my request was denied. The reason given for the denial was a delinquent credit history. I knew my credit was impeccable, and I challenged the decision. The loan officer talked with me and soon realized something was wrong. My wife's name was Kimberley, and at the time of this credit request I didn't have any children. The credit report showed my wife as having a different name, and it named several children. Obviously, the report was inaccurate.

Upon further investigation, it was discovered that the credit bureau had issued the wrong credit report to my bank. My first and last name were the same as the

Financial Statement

Your Company Name
Your Company Address
Your Company Phone Number

Date of statement: _____

Statement prepared by: _____

Assets

Cash on hand	$ 8,543.89
Securities	$ 0.00

Equipment
2004 Ford F-250 pick-up truck	$14,523.00
Pipe rack for truck	$ 250.00
40' Extension ladders (2)	$ 375.00
Hand tools	$ 800.00
Real estate	$ 0.00
Accounts receivable	$ 5,349.36

Total assets	$29,841.25

Liabilities

Equipment
2004 Ford F-250 truck, note payoff	$11,687.92
Accounts payable	$ 1,249.56

Total liabilities	$12,937.48

Net worth	$16,903.77

FIGURE 5-2   A sample financial statement.

person with the poor credit history. However, my middle initial was different, my wife's name was different, and I didn't have any children. It happened that I lived on the same road as this other fellow. It was certainly a strange coincidence, but if I had not questioned the credit report, I would not have been able to build my first home.

I know from first-hand experience that credit reports can be wrong. I have seen various situations in the past when my clients and customers were victimized by incorrect credit reports. If you are turned down for credit, get a copy of your credit report and investigate any discrepancies you discover. And make it a practice to review your credit report periodically. If you find erroneous entries, contact the reporting agency quickly to report the error and have it corrected.

Explanation Letters to Resolve Poor Credit Reports

If your poor credit rating is due to extenuating circumstances, letters of explanation may help. If there was a good reason for your credit problems, a letter that details the circumstances might sway a lender in your direction. Let me give you a true example of how a letter of explanation made a difference to one of my customers.

A young couple wanted me to build a house for them. During the loan-application process it was revealed that the gentleman had allowed his vehicle to be repossessed. This appeared to be a deal-stopping problem. I talked with the young man and learned the details behind the repossession; at my suggestion, he wrote a letter to the loan processor. In less than a week, the matter was resolved and the couple was approved for the new home loan. How did this simple letter change their lives? The man had been a victim of bad advice.

This man had a new truck with high monthly payments. When he decided to get married, he knew he couldn't afford the payments. He went to his banker and explained his situation. The loan officer told the man to return the truck to the bank and the payments would be forgiven. However, the banker never told the man that this act would show up as a repossession on his credit report. My customer returned the truck with the best of intentions, acting on the advice of a bank employee.

When the problem cropped up and the bank employee was contacted,

PRO POINTER

If your poor credit rating is due to extenuating circumstances, letters of explanation may help. If there was a good reason for your credit problems, a letter that details the circumstances might sway a lender in your direction.

he confirmed my customer's story. The mortgage lender for the house evaluated the circumstances and decided that the man was not irresponsible. It was decided that he had acted on the advice of a banking professional, and my customer's loan was approved. If you have an explanation for your credit problems, let your loan officer know about it. Once-in-a-lifetime medical problems could force you into bankruptcy, but they may be forgiven in your loan request. If you provide a detailed accounting that describes your reasons for poor credit, you may find that your loan request will be approved.

Seven Techniques to Assure Your Credit Success

I am about to give you seven techniques to enhance your credit success. This is not to say that these seven methods are the only way to establish credit, but they are proven winners. Let's take a closer look at how you can make your credit desires a reality.

Get a Copy of Your Credit Report

In most cases with a written request you can get a copy of your credit report. This is a wise step to take in establishing new credit. By reviewing your credit report, you can straighten out any incorrect entries before applying for credit. It is better to clean up your credit before a credit manager sees the problems. Even if the report contains false information, a bad impression may already be formed by the credit manager.

Prepare a Credit Package

If you prepare a credit package before applying for credit, the chances of having your credit request approved increase. What should go into your credit package? If you own an existing business, your package should include financial statements, tax returns, your business plan, and all the normal credit information that is typically requested. If you are a new company, provide a strong business plan and the normal credit information. A copy of your personal budget may also be helpful.

Picking the Right Lender

Picking the right lender is a key step in acquiring new credit. Do some homework and find lenders that make the type of loans you want. Once you have your target lenders, take aim and close the deal.

Don't Be Afraid to Start Small

Don't be afraid to start small in your quest for credit. Any open account you can get will help you. Even if you are only given a credit line of $250, that's better than no credit line at all.

Use It or Lose It

When you get a credit account, use it or lose it. Open accounts that are not used will be closed. In addition, an open account that doesn't report activity will do you no good in building your credit rating.

Pay Your Bills on Time

Always pay your bills on time. Having no credit is better than having bad credit. If you have accounts that fall into the past-due category, your credit history will suffer, and you will be plagued by phone calls from people trying to collect your overdue account.

Never Stop

Never stop building your credit standing. The more successful you become, the easier it will be to increase your credit lines. However, don't fall into the trap of getting in over your head. If you abuse your credit privileges, it will not be long before you are in deep financial trouble.

Credit for Your Customers

Have you thought about helping your customers obtain credit? A lot of builders don't take an active interest in helping their customers find financing. Many builders leave loan issues to real-estate brokers and the individuals who are looking to buy a house. If you are willing to participate in setting up easy financing for your customers, you should get more business.

I've worked with financing for a long time, both as a builder and as a broker. It is usually financing that makes or breaks a house deal. Having people who want you to build them a house is not worth much if there is no money available for them to pay you with. It's not your responsibility to find financing for customers, but if you do, you're likely to build more houses.

Permanent Mortgages

Permanent mortgages are available in so many forms that an entire book could be written on them. Traditional 30-year fixed-rate loans and adjustable-rate loans (ARMs) are the two most common types of financing used for homes. Even within these two categories, there can be many differences in terms and conditions. Federally assisted financing is available, as are many other types of loans.

PRO POINTER

As a builder, you should not start construction on a custom home until you know that your customer has a firm commitment for a permanent loan. This type of loan might come from a bank, a savings-and-loan, a credit union, a mortgage banker, or some other source.

As a builder, you should not start construction on a custom home until you know that your customer has a firm commitment for a permanent loan. This type of loan might come from a bank, a savings-and-loan, a credit union, a mortgage banker, or some other source. Your customers are not going to be aware of the many types of loan programs that are available. If the customers are working with a good real-estate broker, the broker can introduce the customers to lenders and loan options. When customers come to you directly without the aid of a broker, you should be able to guide them to some quality lenders. To do this, you must make some inquiries and establish a rapport with lenders before they are needed. Also, find out what your customers will be required to bring to a loan application before the need arises.

Construction Loans

As you interview lenders, try to find ones who will approve your customers for permanent financing and a construction loan. Your risk is reduced if the customers sign for the construction loan in their own names. When a builder borrows money on a construction loan, there are two closings. The builder has to close on the land and put the property in the name of the

PRO POINTER

As you interview lenders, try to find ones who will approve your customers for permanent financing and a construction loan. Your risk is reduced if the customers sign for the construction loan in their own names.

building business; when the house is completed, it is sold to the customer, and another closing is needed. This increases the cost of the home. By having your customers set the construction loan up in their name, you reduce your risk and their cost.

While you are shopping around for sources of construction loans, you should ask about terms, conditions, and draw disbursements. Some construction loans are active for six months, others run for nine months, and some go for a full year. Make sure that the term of the loan is adequate for you to complete the house.

What is the loan-to-value ratio of the construction loan? A 70-percent loan is common. Loans with a 75-percent loan ratio can frequently be found, and some lenders still loan up to 80 percent of a property's appraised value. The higher the loan-to-value ratio is, the better off you are.

Does the construction loan provide a land-acquisition disbursement? Most construction lenders will advance money from a construction loan to buy a building lot. This can become a big issue. If a lender won't advance money for land acquisition, you or your customer will have to come up with a lot of cash on your own. Find a bank that will front the money for land.

The interest rate on a construction loan is normally higher than the rates quoted for permanent financing. This is because construction loans have short terms and are a higher risk than long-term mortgages. You need to know what the interest rate will be so that you can calculate the cost of financing into your house price, unless your customers will be paying the interest payments out of their own funds.

The cost of points and closing costs can add thousands of dollars to the cost of a house. Each point charge is equal to 1 percent of the loan amount. In other words, a construction loan that charges one point for a $100,000 loan will cost $1,000 plus closing costs. Determine what all the financing costs will be before you price a job, and decide how the money is

PRO POINTER

As you interview lenders, try to find ones who will approve your customers for permanent financing and a construction loan. Your risk is reduced if the customers sign for the construction loan in their own names. When a builder borrows money on a construction loan, there are two closings. The builder has to close on the land and put the property in the name of the building business; when the house is completed, it is sold to the customer, and another closing is needed. This increases the cost of the home. By having your customers set the construction loan up in their name, you reduce your risk and their cost.

going to be paid. Are you going to pay it and add the expense onto the price of a house, or will your customers pay all financing fees privately and then pay you for the construction work? Construction loans are a key element in building most houses, so spend some time talking to lenders and get the facts on what types of financing are available for you and your customers.

Bridge Loans

Bridge loans can be used to overcome a common problem. Many people have homes that they must sell before they can buy a new one. Sometimes home-buyers have a contract to sell their house and are waiting for the closing to take place. If either of these types of people wants you to build a new house for them, they may think that they have to wait until their deals close to start construction. This isn't always true. Sometimes a bridge loan can be used to get the ball rolling on a new house while an old house is being sold or closed.

PRO POINTER

If your customers are strong enough financially, they should be able to arrange a bridge loan that will allow them to have a new house built while they are making arrangements to leave their existing home. If you and your customers are not aware of this option, you could lose months of production time waiting for one deal to close so that you can start a new deal. Check into bridge loans when you talk with lenders and be prepared to share your knowledge with customers; it could get you more work.

If your customers are strong enough financially, they should be able to arrange a bridge loan that will allow them to have a new house built while they are making arrangements to leave their existing home. If you and your customers are not aware of this option, you could lose months of production time waiting for one deal to close so that you can start a new deal. Check into bridge loans when you talk with lenders and be prepared to share your knowledge with customers; it could get you more work.

An Edge

Builders who are able to help customers set up financing have an edge on their competition. Knowing what customers can do and where they can do it makes you a more formidable adversary to other builders. When customers are shopping for builders, the contractor who can make the job and the financing easy and comfortable is likely to win the bid. Don't underestimate the importance of financing. Once you become acquainted with loans and lenders, you should find that sales are easier to make.

Bonds

The Association of General Contractors has estimated that half of all contracting companies go out of business within six years, and both bankers and owners are aware of situations in which a builder went out of business and left the bank or the owner high and dry. Bonds offer protection against builder default, and the ability of a contractor to be "bonded" adds to his reputation for integrity and provides a prospective lender or client with a clear indication of the contractor's financial stability.

As part of your education in financial matters, the subject of bonds should be discussed. Bonds differ from insurance in that insurance protects you from the unknown but bonds ensure that funds will be available to complete a project if the contractor fails to pay for all labor and material (payment bond) or fail to "perform" (performance bond) per the conditions of the contract.

The two primary types of bonds that you as a builder are likely to encounter are the payment bond and the performance bond. The payment bond guarantees that the builder will pay all subcontractors, workers, and material and equipment suppliers for a particular job. A performance bond protects the owner from any financial loss should the builder fail to fulfill the terms and conditions of the contract.

Some owners and some banks may insist that a builder provide payment and performance bonds before he or she can receive a bank loan or, in the case of an owner, before they will enter into a contract.

Although you may never experience a request for a bond, as part of

PRO POINTER

The Association of General Contractors has estimated that half of all contracting companies go out of business within six years, and both bankers and owners are aware of situations in which a builder went out of business and left the bank or the owner high and dry. Bonds offer protection against builder default, and the ability of a contractor to be "bonded" adds to his reputation for integrity and provides a prospective lender or client with a clear indication of the contractor's financial stability.

your education as a builder, you ought to be aware of them—what they do and how to go about getting "bonded."

Bonds are sold through insurance companies, and the process of obtaining a bond is often as important as the bond itself. Bonding companies will only issue bonds to contractors if they are financially sound and have good management practices in place. Many contractors use the bonding process to evaluate their company from top to bottom and take advice from the bonding agent on how to improve their financial and management strengths. A typical submission to a bonding agent will include:

- An organizational chart of your company
- Detailed resumes of you and your key employees
- A business plan outlining your type of work, how you plan to grow, and what your profit goals are
- A list of your largest and latest projects
- References from some of your subcontractors and suppliers
- Evidence of a line of credit from your bank
- Letters of recommendations from previous owners
- Various statements from your accountant: a balance sheet showing assets, liabilities, and net worth; an income statement; a statement of cash flow; a schedule of contracts in progress and contracts completed; and a schedule of general expenses to show your overhead
- A year-end audited financial statement

Even if you never have to obtain a bond, the steps necessary to be bonded can provide you with some guidelines you may need in the future to build financial strength as your company continues to grow.

Working from Home versus Setting Up an Office

Office space can be a major factor in the successful operation of your business. There is little need for a construction business to have an office in the high-rent district; most contracting businesses can function from a low-profile location. You can even work from home in some cases. This is not to say that office space is not needed or is not important. Whether you work from your home or a penthouse suite, your office has to be functional and efficient if you want to make more money.

Should You Work from Home or from a Rented Space?

Whether to work from home or from a rented office can be a very difficult question to answer, but with some thought and my experience and advice you can make an intelligent decision. I have worked from home and from commercial offices, and my experience has shown that the decision to rent commercial space is dictated by your own self-discipline, the type of business you are running, and cost. Let's explore the factors you should consider when thinking about where to set up shop.

Self-Discipline

Self discipline is paramount to your success in business, and the ability to work out of an office in your home is a good test. It is easy to spend too much time around the breakfast table or go for a stroll around your farm. Working from home is very enjoyable, but you do have to set rules and stick to them. Designating a specific room for your home office, hopefully away from daily household activity, is a start in the right direction.

Storefront Requirements

Some businesses require storefront exposure, but builders rarely have such needs. Other trades, such as plumbers or electricians, may need to display the fixtures they sell, but most builders have no need to do this. They can send their customers to supplier's showrooms or show them catalogs of plumbing and electrical fixtures, kitchen cabinets, windows, doors, and hardware selections. Having an office that your customers can visit helps to build professionalism, and this is an important element for success. But your first step is looking at your finances—can you afford the expense of a rented office?

PRO POINTER

Self discipline is paramount to your success in business, and the ability to work out of an office in your home is a good test. It is easy to spend too much time around the breakfast table or go for a stroll around your farm. Working from home is very enjoyable, but you do have to set rules and stick to them. Designating a specific room for your home office, hopefully away from daily household activity, is a start in the right direction.

Home Office

A home office is a dream of many people. Putting your office in your home is a good way to save money if it doesn't cost you more than you save. Home offices can have a detrimental effect on your business. Some people will assume that if you work from home you are not well established and may be a risky choice as a contractor. Of course, working from home doesn't mean your business is in financial trouble, but some customers are not comfortable with a company that doesn't have a commercial office space.

I work from home now, and I have worked from home at different times for nearly 20 years. I love it. I am also disciplined in my work ethic, and even though my office is in the home, it is professionally set up. When clients come to my home office, it is obvious that I am a professional. I will talk more about setting up a home office a little later, but remember to take your home office seriously. Having an office in your home allows you to write off certain household expenses. By calculating the size of your office as a percentage of the total square footage of your home, you can deduct a certain portion of utility costs, real-estate taxes, and mortgage payments. All costs to furnish your office should be tax-deductible as well. So if you are considering setting up an office in your home, you ought to talk to an accountant and find out the tax advantages and tax rules pertaining to a home office.

Commercial Image

A commercial office can give you a commercial image. This image can do a world of good for your business. However, commercial office space can be expensive and add greatly to your overhead expense. Before you jump into an expensive office suite, consider all aspects of your decision. We will talk more about the pros and cons of commercial offices as we continue with the chapter.

Assess Your Office Needs

Before you can decide on where to put your office, you need to assess your office needs. This part of your business planning is important to the success of your business. Can you imagine opening your business in an expensive office and then, say six months later, having to move out because these expenses are hurting you? Not only would that situation be potentially embarrassing, it would be bad for business. Once people get to know your office location, they expect it to stay there or to move up. A downward move, like the scenario described, would concern people, who may think your business is not doing well, and scare off a percentage of future customers.

How Much Space Do You Need?

One of the first considerations in choosing an office is how much space you need. If you are the only person in the business, you may not need a lot of office space. When you consider your space requirements, take the time to sketch out your proposed office space. It helps if you make the drawing to scale. You can also make some templates, to scale, of a desk, chair, filing cabinets, storage shelves, and a table for a copier. Move them around in the space you think you need for an office to determine whether you need more or less than you originally thought. And remember to leave enough room on the scaled drawing to allow your chair to be pushed back from your desk and not hit any other furniture. How many people will you meet with at any given time? How many desks will be in the office? I now have only a three-person office. This does not count subcontractors who have their space off-site. My office is a separate building on my land. It is not in my home, but it is right across the driveway. This makes it easy to define work space. When I have worked out of rooms in the home, the discipline needed to be successful is considerable.

I have two desks and a sorting table in the primary office. In addition, I have another room for files, supplies, the copier, and similar items. There is another space set up as a photography studio. My storage buildings contain my tools, equipment, and supplies. Even a small business can have a need for large spaces.

PRO POINTER

When you are designing your office, consider all your needs. Desks and chairs are only the beginning. Will you have a separate computer workstation? Do you need a conference table? Where will your filing cabinets go? Where and how will you store your office supplies? How many electrical outlets will be needed? The more questions you ask and answer before making an office commitment, the better your chances are of making a good decision.

When you are designing your office, consider all your needs. Desks and chairs are only the beginning. Will you have a separate computer workstation? Do you need a conference table? Where will your filing cabinets go? Where and how will you store your office supplies? How many electrical outlets will be needed? The more questions you ask and answer before making an office commitment, the better your chances are of making a good decision.

Do You Need Commercial Visibility?

Do you need commercial visibility? Almost any business can do better with commercial visibility, but the benefits of this visibility may not warrant the extra cost. If you are out in the field working every day and you don't have an employee in the office, what good will it do you to have storefront exposure? If your business allows you to remain in the office most of the time, a storefront might be beneficial. You would get some walk-in business that you wouldn't get working out of your home. Another advantage of commercial space is that it builds credibility for your company in the eyes of customers.

Do You Need Warehouse Space?

If you have to keep large quantities of supplies on hand or deal in bulky items, warehouse storage may be essential. One solution for office and storage space is a unit that combines both under the same roof. These office/warehouse spaces are efficient, professional, and normally not very expensive if you can find them.

If you don't need access to the materials in storage, renting a space at a private storage facility can be the best financial solution to your needs. At one time, I ran my company from a small office and a private storage facility. This arrangement was not convenient, but it was cost-effective, and it worked. Once you have evaluated your needs, you can consider where to put your office.

Location Can Make a Difference

In business, location can make a difference. If you cater to people in the city, living in the country can be inconvenient and cause you to lose customers. Having an office where you are allowed to erect a large sign is excellent advertising and builds name recognition. If your office is in a remote section of the city, people may find it difficult to get to your office and stay away. If you work from your home and your home happens to be out in the country, customers may not be able to find you even if they are willing to try. Location is an important business decision.

How will your office location affect your public image? Public opinion is fickle, and if your business is perceived to be successful, it probably will become successful. On the other hand, if the public sees your business as a loser, look out. It is unfortunate that we have to make some decisions and take some actions just to create a public image, but there are times when we must.

What does the location of your office have to do with the quality of your business? It probably has nothing to do with it, but the public thinks it does. For this reason, you must direct your efforts to the people you hope will become your customers.

Aside from the prestige of an office location, you must consider the convenience of your customers. If your office

PRO POINTER

In business, location can make a difference. If you cater to people in the city, living in the country can be inconvenient and cause you to lose customers. Having an office where you are allowed to erect a large sign is excellent advertising and builds name recognition. If your office is in a remote section of the city, people may find it difficult to get to your office and stay away. If you work from your home and your home happens to be out in the country, customers may not be able to find you even if they are willing to try.

is at the top of six flights of steps with no elevators available, people may not want to do business with you. If there is not adequate parking in the immediate vicinity of your office, you could lose potential customers. All these factors play a part in your public image and success.

How Much Office Can You Afford?

This would appear to be a simple question, but answering it may be more difficult than you think. When you look at your budget for office expenses, you must consider all the costs related to the office. These costs might include heat, electricity, cleaning, parking, snow removal, and other similar expenses. These incidental expenses could add up to more than the cost to rent the office.

If you rent an office in the summer, you might not think to ask about heating expenses. In Maine, the cost of heating an office can easily exceed the monthly rent on the space. When you prepare your office budget, take all related expenses into account and arrive at a budget number that you are comfortable with and that is realistic.

PRO POINTER

Before you rent an office, consider ups and downs in your business cycle. If you are in a business that drops off in the winter, will you still be able to afford the office? Do you have to sign a long-term lease, or will you be on a month-to-month basis? It usually costs more to be on a month-to-month basis, but for a new business it might be well worth the extra cost.

When you begin shopping for an office outside your home, ask questions and lots of them. Who pays for trash removal? Who pays the water and sewer bill? Who pays the taxes on the building? These questions are important because some leases require you, the lessee, to pay the property taxes. Who pays the heating expenses? Who pays for electricity? Who pays for cleaning the office? Does the receptionist in the lower level of the building cost extra? If there is office equipment in a common area, such as a copier, for example, what does it cost to use the equipment?

Ask all the questions and get answers. If you are required to pay for expenses such as heat or electricity, ask to see the bills for the last year. These bills will give you an idea of what your additional office expenses will be.

Before you rent an office, consider ups and downs in your business cycle. If you are in a business that drops off in the winter, will you still be able to afford the office? Do you have to sign a long-term lease, or will you be on a month-to-month basis? It usually costs more to be on a month-to-month basis, but for a new business it might be well worth the extra cost. If you sign a long lease and default on it, your credit rating can be tarnished. What happens if your business booms and you need to add office help? If you are in a tiny office with a long-term lease, you've got a problem. If you do opt for a long-term lease, negotiate for a sublease clause that will allow you to rent the office to someone else if you have to move.

It can be easy to dream about how a new office will bring you more business, but don't put yourself in a trap. When you project your office budget, base your forecast on your present workload. Better yet, if you've been in business awhile, base the projections on your worst quarter for the last year. If you can afford the office space in the bad times, sign the lease. If you can only afford it during the summer boom, you're probably better off without the office.

Many new business owners may look for space with marble columns, fancy floors, wet bars, and all the glitter depicted in offices on television. Unless you are independently wealthy, these high-priced work spaces can rob you of profits and put you out of business.

Potential customers may view a fancy office as an expense that is tacked on to the price of the house you are selling. Don't assume that an expensive office is going to return higher net earnings.

Answering Services Compared to Answering Machines

When answering services are compared to answering machines, you may find many different opinions. Most people prefer to talk a live person rather than an electronic device. However, as our lives become more automated, the public is slowly accepting the use of electronic message storage and retrieval.

When you shop for services, which do you prefer, an answering service or an answering machine? Do most of your competitors use machines or people to answer their phones? This is easy to research; just call your competitors and see how the phone is answered. The use of an answering machine may cause you to lose business, but that doesn't mean necessarily that you should not consider it.

Answering machines are relatively inexpensive, and most machines are dependable. These two points give the answering machine an advantage over an answering service. Answering services are not cheap, and they are not always dependable. Answering services can page you to give you important and time-sensitive messages; answering machines can't. Another point for answering services (and most answering machines) is that you can access them from your mobile phone and pick up your messages while out of the office.

To determine which type of phone answering method you should use, make a list of the advantages and disadvantages of each. Once you have your list, arrive at a first impression on which

PRO POINTER

When answering services are compared to answering machines, you may find many different opinions. Most people prefer to talk a live person rather than an electronic device. However, as our lives become more automated, the public is slowly accepting the use of electronic message storage and retrieval.

option you should choose. It may be necessary to change your decision later, but at least you will be off to a reasonable start.

There are several qualities you should look for in an answering machine. It should include the ability to check your messages remotely. Most modern answering machines can be checked for messages from any phone. Choose a machine that allows the caller to leave a long message. Many machines will allow the callers to talk for as long as they want. These machines are voice-activated and will cut off only when the caller stops speaking. Pick a machine that will allow you to record and use a personal outgoing message. Some of the answering machines are set up with a standard message that you can't change. It will be beneficial to customize your outgoing message. If you buy an answering machine that meets all these criteria, you should be satisfied with its performance.

In considering answering services, price is always a factor, but don't be guided solely by price; you get what you pay for. Find an answering service that answers the phone and takes messages in a professional manner. You want a dependable service, one that will deliver messages to you in a timely manner. Ask if you can provide a script for the operators to use when answering your phone. Some services answer all the phones with the same greeting, but many will answer with your personalized message.

Ask about the hours of the day for your coverage. Most services provide 24-hour service, but that generally costs more. Ask if the service will page you for time-sensitive calls; most will. Determine if your bill will be a flat rate or if it will fluctuate, depending on the number of calls you get. Inquire about the length of time you must commit to the service. Some answering services will allow you to go on a month-to-month basis, and others want a long-term commitment.

If you decide to use an answering service, check on the performance periodically. Most services provide a special number for you to call to pick up your messages, especially if they base your bill on the number of calls received. Even if you will have to pay for calling in on your own line, do it every now and then. When you call in on your business number, you can get first-hand proof of how the operators handle your calls. Have friends call and leave messages. The operators won't recognize the voices of these people and will treat them like any other customer. This is the best way to check the performance of human answering services.

Machines or humans? I think the business you lose with an answering machine is more valuable than the money you save. If you can hire a personalized answering service, you should. I have tried having my phones answered each way, and I am convinced that human answering services are the best way to go.

Pulling together all the components of a functional office can take time. Don't expect to make the best decisions on your first attempt. You may have to experiment to find out what works and what doesn't. If you remember just one thing, it is this: Don't jump into a long lease; you are better off growing slowly than not growing at all.

PRO POINTER

Machines or humans? I think the business you lose with an answering machine is more valuable than the money you save. If you can hire a personalized answering service, you should. I have tried having my phones answered each way, and I am convinced that human answering services are the best way to go.

Building on Speculation—the Risks and the Rewards

Building houses on speculation is more risky than it used to be. There was a time when builders could sell spec houses faster than they could build them. Then times changed, and builders found themselves holding spec houses for a year or more, using up all their profit by paying interest payments on their construction loans. Opportunities for spec builders are not as good as they once were.

We saw a surge in the building market when interest rates dipped to extreme lows. The rates are still favorable, but the market is flooded with homes in many locations. There are some regions that can absorb more new housing, but there are plenty of places where selling a house is a major headache. The subprime loans tend to be blamed. Surely, this is part of the situation. The boom in building has made the market difficult to play in safely.

What are you to do? Building will continue. Sustainable building is growing. But should you build a spec house? Are you willing to take the risk? Is the potential reward worth the risk? It is a tough question to answer.

Can you be a builder without gambling on spec houses? Yes, you absolutely can. I have done it for years, and so have many other builders. What are you going to do? If you don't know, this chapter should help you to come to a viable decision.

Builders who concentrate on spec houses avoid a lot of the hassles that custom builders must face. When you build on spec, you call the shots until a buyer comes along. Custom builders have to work under the watchful eye of their customers from start to finish. It is considerably easier to sell a house that has already been built than it is to sell one from blueprints. Spec builders enjoy the luxury of selling tangible products, while custom builders are often selling paper and blue lines. The list of advantages of building on speculation goes on, and we will talk more about them as we progress into this chapter. On the other hand, builders who sell from plans also have advantages. For example, your customers can customize a home that only exists on paper much more easily than a house under construction. There are pros and cons to both methods of becoming a successful contractor.

Picking Lots and Plans

Picking building lots and house plans is one of the first things that a spec builder has to do. If you're going to build on spec, you have to have lots to build on. Don't take this part of your job lightly. The lot location, size, shape, and adjoining and adjacent properties contribute not only to its market value but also to its appeal to customers.

Speculative building is a gamble in every way. You are buying a lot that you hope will make some homebuyer happy. Then you are building a house that you feel will have market appeal. If you're able to sell the house, you make money. When a spec house stagnates

PRO POINTER

Speculative building is a gamble in every way. You are buying a lot that you hope will make some homebuyer happy. Then you are building a house that you feel will have market appeal. If you're able to sell the house, you make money. When a spec house stagnates and doesn't sell, you have to buy it yourself or give it back to the bank, which will damage your credit and your credibility for a long time to come.

and doesn't sell, you have to buy it yourself or give it back to the bank, which will damage your credit and your credibility for a long time to come. You could put $50,000 profit in your pocket, or you could wind up in a bankruptcy court. There is, of course, some middle ground with different options. The bottom line is this: Spec building is a gamble.

Good spec builders have a feel for their market. They stay in touch with brokers, other builders, current subdivision projects, and, of course, competitive advertising. The best builders track comparable sales by working with either appraisers or brokers. By doing this a builder can see exactly which types of houses are selling, the price they are selling for, and how long it is taking to sell them. This type of information is invalu-

FIGURE 7-1   Aesthetic design can sell houses faster (courtesy of ECO-Block, LLC).

able when deciding what type of houses to build on speculation. You can't take the gamble out of spec building, but you can pad the odds in your favor with enough research.

If you have access to a book of comparable sales or an online comparative sales analysis—and any real-estate broker worth a second look will have this type of access— you can get a wealth of information to help you in choosing plans for spec houses. A comp report will list the foundation size of a home, the size and number of rooms in the house, the type of electrical service provided, specifications on the heating system, and even the floor coverings. By reading through comp reports, you can see exactly what's going on in your real-estate market, and the information is generally accurate and dependable.

Watching competitive advertisements is a necessary function, but you have to take the ads with a grain of salt. Many builders and brokers are good at running teaser ads that don't tell you much until you come in for a qualifying meeting. Unless you pose as a prospective buyer, you will gather only limited information from ads. Read them. Pay attention to them. Learn from them. But don't rely on them.

Ride around new subdivisions and see what other builders are building. Inspect the construction, if you can, to see how it stacks up to what you will offer your customers. Compile as much data as you can before you commit to building a certain type of house in a particular way. I believe the book of comparable sales is your best tool, but ads and ride-bys are effective in their own right.

Once you have as much information as you can get, you have to start making decisions. You can run with the pack, or you can break away and build something completely different. Getting radical is risky. It is normally better to stick with what's working for other builders but with some special twist of your own. Picking the right plan involves more than simply saying that you are going to build a split-foyer or a ranch. You have to assess and target your market.

Target Your Market

If you target your market for spec houses, you are much less likely to fail in your attempt to sell your houses quickly. Some markets overlap. Others don't. A lot of builders want to build large, expensive homes because their profit is usually established in proportion to the home's value. But some builders deal in less expensive homes and operate on a volume basis. Both approaches have merit.

Builders of massive homes will boast that they only have to sell one or

PRO POINTER

If you target your market for spec houses, you are much less likely to fail in your attempt to sell your houses quickly. Some markets overlap. Others don't. A lot of builders want to build large, expensive homes because their profit is usually established in proportion to the home's value. But some builders deal in less expensive homes and operate on a volume basis. Both approaches have merit.

two a year to make a comfortable living. A builder who specializes in starter homes might have to build two or three times as many homes to make the same money. But if a massive house doesn't sell, the carrying cost is substantial, and no income is derived. If I build six small houses and sell four of them, I can afford to carry the other two and still survive. Building one or two big houses and hanging your hat only on those pegs could be disastrous. On the other hand, I do have to sell more houses, and this could be a problem.

FIGURE 7-2 Speculative building of multi-family buildings requires the use of amenities that residents will appreciate (courtesy of ECO-Block, LLC).

Buyers of expensive homes are usually well qualified, but they often have a house to sell. Many of my buyers are moving out of rental property, so I can put deals together quickly that are not contingent on other sales. This, in my opinion, is a big advantage. I might have to work a little harder to find a loan for my first-time buyers, but if my connections come through, it's a done deal.

You have to decide what price range you will build in. I can't tell you what will work best for you. First-time homes have done well for me, but I've built many more expensive houses with good success. The key is to know what you want to do in advance. You can't pick a plan and run with it until you have all the details worked out.

FIGURE 7-3   Good natural lighting and spacious rooms sell houses (courtesy of Natural Cork and More).

Hitting Your Mark

Hitting your mark, once you have targeted your market, is easy if you know what to do. If you have defined your market tightly enough, reaching that market with the tools available to you from today's options is simple. Don't attempt to advertise until you feel quite comfortable that your target audience has been identified in a way that is easy to reach.

Good advertising is a bargain. Poor advertising can be a disaster. Knowing how to maximize your advertising dollar is a critical aspect of being successful as a spec builder. If you blindly spend hundreds of dollars each week for display advertising in your local paper, you could be wasting your hard-earned profit. Potential homebuyers often read ads in newspapers, but these prospects are not always the best ones to work with. There is often a hidden market that is much richer to mine.

PRO POINTER

Good advertising is a bargain. Poor advertising can be a disaster. Knowing how to maximize your advertising dollar is a critical aspect of being successful as a spec builder. If you blindly spend hundreds of dollars each week for display advertising in your local paper, you could be wasting your hard-earned profit. Potential homebuyers often read ads in newspapers, but these prospects are not always the best ones to work with. There is often a hidden market that is much richer to mine.

General advertising to the public works, but targeted marketing is often more cost-effective. Running commercials on television is a great way to build name recognition, but it probably isn't an efficient way to sell one or two spec houses. If you are promoting an entire development, television advertising is great, but it costs too much to use for the sale of a single spec house.

When you buy advertising on radio, television, or in the newspaper, the cost of your ad is directly related to the number of people who listen to, watch, or read your chosen medium. Your ads could be reaching tens of thousands of people, but how many of those people want what you have to sell? You simply don't know. You can gauge it to some extent with demographics, such as age groups of readers and listeners, but that's a far cry from a tightly defined audience.

What do you think is the best way to sell one or two spec houses? Most builders would either say that listing the homes with a broker or advertising the houses for sale in the local paper will work. Either of these methods can work, but I've found great success with other methods. Let me give you a couple of examples.

Let's say that you are building a name for yourself as a builder of log homes. On a percentage basis, the demand for log homes is not too great. But there are plenty of people who will buy log homes. It's just that their numbers are minuscule when compared to the general home-buying public. With this being the case, you would be paying a tremendous amount of money for advertising on a per-lead basis if you advertised in general media. Your ad bill will be based on the total circulation of a newspaper, not on the number of readers who might be interested in a new log home. Is there a better way? Yes, there is.

Direct mail is an awesome way of reaching a target market. By renting name lists, you can mail promotional pieces to people who fit the criteria you are looking for in a buyer. What, for example, would be points of interest in your selection of a name list for log-home buyers? Income would certainly be one qualifier. If the house you are building is small or large, the number of members in a family might be important in culling your name list. A family of five would not be likely to buy your two-bedroom cabin. Have the prospects shown any past interest in log homes? If they subscribe to magazines about log homes, and a mailing-list company can probably tell you if they do, your potential customers are a prime target.

PRO POINTER

Direct mail is an awesome way of reaching a target market. By renting name lists, you can mail promotional pieces to people who fit the criteria you are looking for in a buyer.

When you rent names from reputable list brokers, you can require that all sorts of criteria be met by the names placed on your list. Each set of criteria adds to the cost of a list, but the breakdowns are well worth their cost. The list of names that I'm preparing to order soon is going to cost me 6 cents per name. My list will arrive on pressure-sensitive labels, and it is sorted by location, income, and the fact that everyone on it lives in rental property. Prices for these lists can offer quite a bargain.

The list I'm ordering will be used to solicit first-time homebuyers. That's why I wanted to make sure that everyone on the list lived in rental property. The people will not be encumbered by a home to sell, so I should be able to make some quick deals.

Direct mail is not the only way to hit your target. Most towns and cities have publications that are aimed at particular markets. There is a military base located near where I work. The base is a good source of incoming housing prospects. By advertising in the small paper that serves the base, I can target my ads to buyers who have an ability to use VA loans.

If I wanted to hit the out-of-state market that is so lucrative here in Maine, I know of a few national magazines that I can advertise in that have proven successful as real-estate vehicles. Vacation homes and camps are very popular in Maine. I could advertise in many types of local outdoor publications to get work building seasonal cottages and cabins.

The point I'm making is this: Once you know whom you want to sell to, there are many good ways to get your message through to them. Don't waste your money on expensive general media campaigns when you are selling a single spec house. An ad in the paper is a good idea, but it doesn't have to be a big, expensive ad. The money spent on advertising is one of your largest overhead expenses, so invest it wisely.

PRO POINTER

The point I'm making is this: Once you know whom you want to sell to, there are many good ways to get your message through to them. Don't waste your money on expensive general media campaigns when you are selling a single spec house. An ad in the paper is a good idea, but it doesn't have to be a big, expensive ad. The money spent on advertising is one of your largest overhead expenses, so invest it wisely.

A Safety Net

Would you walk a tightrope over a pool of hungry crocodiles without a safety net below you? I wouldn't do it under any conditions, but there are always some daredevils out there. Building houses on speculation is not the same as balancing yourself on a cable over gaping crocodile mouths, but it can be nearly as dangerous. If you hope to survive year after year as a spec builder, you have to plan for the unwanted experiences that are likely to surface from time to time.

I haven't built a spec house in several years. My reasoning is simple: I've had enough contract work to do so that I don't have to take the risk of building on speculation. There are still times when I would prefer the fewer hassles that go along with spec building, but it's safer building contract houses.

At my peak of spec building, I had an unusual but very effective safety net. I established groups of investors who would buy any spec house that I had trouble selling. The sales to investors had to be discounted deeply, but a 5-percent profit was better than a loss.

During my early years as a spec builder, my safety net was building each spec house with the intent to live in it, at least for a few months. If the house didn't sell quickly, I lived in it until it did. Another ploy I used was to arrange permanent financing for my spec houses to be used as rental property if they didn't sell well. If a house was costing me too much money in interest to keep it on the open market, I would close on the 30-year loan and rent the house out.

You don't have to have a contingency plan when you build on speculation, but you should. A lot can go wrong, and it doesn't take long for a builder to lose all profit in a house to the carrying costs. I strongly suggest that you have at least one backup plan in mind for every spec house you build. Otherwise, your ticket to fame and fortune could wind up being a one-way trip to financial disaster.

Colors and Products

The colors and products associated with a spec house often have to be chosen by the builder. There is risk to this. Colors that you like might not appeal to potential buyers. A brand of carpeting that appeals to you might not suit a buyer. Avoid making these types of decisions until the last minute. On the items that you must choose, make sure that you are taking a neutral route.

The color of a home's roof and siding must be decided on early in the construction stage. Interior paint colors might be able to wait until a buyer is found, but spec builders often choose them. Go with noncommittal colors, such as white, off-white, cream colors, and so forth. Don't pick standout colors that make definitive statements. By the same token, be careful about letting customers choose their colors before their financing is arranged.

Let's say that your spec house is almost ready for interior paint and you have a couple that wants to buy the house. The people want one bedroom painted pink, one papered with dinosaurs, and other rooms painted in a variety of special colors. If you go along

PRO POINTER

The colors and products associated with a spec house often have to be chosen by the builder. There is risk to this. Colors that you like might not appeal to potential buyers. A brand of carpeting that appeals to you might not suit a buyer. Avoid making these types of decisions until the last minute. On the items that you must choose, make sure that you are taking a neutral route.

FIGURE 7-4 Sustainable flooring can provide excellent formal functions (courtesy of Natural Cork and More).

FIGURE 7-5 Kitchens are a key element when building on speculation (courtesy of Natural Cork and More).

with their wishes and paint and paper the house while their loan is being processed, you could wind up in trouble. Suppose their loan request is denied? What will you do with a spec house that is haunted with custom colors? It will be more difficult to sell. It's good to let customers make their own choices on colors, floor coverings, plumbing fixtures, and other items, but you have to make sure that the people you are selling to will in fact be able to complete their obligation to the sale.

My rule-of-thumb as a spec builder has always been to let people pick their own selections and to stop construction until their loan is approved. This slows down the process, and I've made some exceptions to my rule, but generally I've stuck with it. Customers won't like the fact that you will not continue construction until their loan is approved, but you have to weigh the risk. If the customers select colors and products that you feel would be acceptable to most prospective buyers, keep the construction rolling. But when you get someone who has unique taste, or in other words someone who might be considered weird, hold off on the customization until you know the sale is a done deal.

Selling without Having a House to Show

Selling houses from a set of blueprints and some photographs is a skill that takes some time to perfect. Many people are reluctant to buy a house that has not yet been built. But a good number of people relish the thought of arranging the construction of a custom-built home. These are the people you want to talk to. If you can capture their attention, you can make some quick sales without the aid of a model home.

I've had a lot of experience selling homes, both as a builder and as a broker. When I started selling houses with nothing but a set of blueprints to work with, the job was referred to as "selling houses in the dirt." What I was selling was a building lot and a dream. I struggled through my first several attempts to sell a house from plans, and I was not successful. However, I didn't give up, and eventually I made my first sale.

As my sales career grew along with my building business, I made mental notes of what worked and what didn't when I was attempting to make a sale. By keeping records of my activity, I was able to review what I'd done and improve on it. Pretty soon I was quite proficient in the art of selling a set of blueprints and a new home.

In order to presell houses on a regular basis, you are going to have to invest some time in learning how to do it. Basic sales skills are needed, but they are not enough to get the job done. Lots of salespeople can sell well, but few of them can sell dream homes from paper consistently. If you show enough people blueprints and present enough proposals, you will eventually make some sales. But you will waste a lot of time and money in the process. You need to make every sales meeting count.

FIGURE 7-6   Interior design elements, such as arches, provide eye-appeal that can sell a house quickly (courtesy of FrameGuard® Mold-Resistant Wood).

The Basics

There are many books on the market that teach the basics of salesmanship, and you should read some of them. Knowing what to say, how to say it, and when to say it is a big part of the sales game. More important, sometimes, is knowing when to stop talking and start listening. We'll talk more about this in a moment.

We won't go into a full explanation of selling techniques, and I'm probably not enough of an expert in general sales to teach you all that can be learned, but get your hands on some books that deal with nothing but sales skills and read them. Make notes as you go along. Compare what the different authors are telling you. You should start to see a pattern develop as you read the books. The principles behind successful selling are basically the same, whether you are selling insurance, home improvements, or houses.

Once you have gained an awareness of how to use traditional sales skills to increase your business, you can refine the techniques and customize them to suit your personal needs and desires. In other words, you can start with generic sales methods and shape them to the sale of new houses. Some of your customization will come from trial-and-error experiences. Each time you miss a sale, you should learn from the experience. And every time you make a sale, you should recount the strategy that made you successful. To get you started in selling houses from blueprints, I can share with you what I've learned over the last several years.

Talk Less and Listen More

When I'm training new salespeople, I often tell them to talk less, listen more often, and sell. Many sales professionals talk too much and don't listen to what their customers want, and they lose sales because of this. Think about the last time you bought a car or truck. Did the salesperson descend on you and start chatting about insignificant matters? How long did the salesperson monopolize the conversation before letting you get a word in edgewise? Was there a point during the meeting that you wished the talkative salesperson would just go away?

I just bought a new Jeep, and the salesman almost lost the sale because he talked so much. When I went to the car lot, I knew I was going to buy a new Cherokee. There was no doubt in my mind about what I wanted. As I was looking at the selection of vehicles, a salesman approached me and started talking. One of the first things he did was to try to steer me from a new Cherokee to a used one. I didn't want a used one, and I told him this. But he kept pushing the old Jeep on me. It wouldn't have taken much to make me move to another dealer. How-

PRO POINTER

When I'm training new salespeople, I often tell them to talk less, listen more often, and sell. Many sales professionals talk too much and don't listen to what their customers want, and they lose sales because of this.

ever, I simply told him that I was going to buy a new vehicle and that I'd appreciate it if he would let me shop in peace. He went away and I bought the Jeep.

Over the years, I've trained a lot of salespeople. Some of them have been groomed to sell home improvements for my remodeling business. Others have been taught to sell new houses, both built and in the dirt. As part of my training course, I sit in on sales meetings with the new recruits when they are first getting started. It never ceases to amaze me how often the people blow deals by talking when they should be listening.

It's necessary for you to talk to customers, but you have to know when to talk and when to listen. A little chitchat to break the ice is needed. You must be well versed in your product knowledge and be able to weave in interesting and pertinent facts as your sales meeting progresses. But in many cases, you have to keep quiet and let your customers explain what it is that they want.

If you bore your customers with a canned pitch, like a lot of salespeople do, you may alienate them and lose your sale. Sitting at a table telling a young couple how great an expandable Cape Cod is for a starter home could come across as an insult. For all you know, the couple may be wealthy and looking for a six-bedroom colonial to fill with children. If you hammer them with your cheap Cape, the meeting could turn sour quickly.

People Like to Talk

Most people like to talk about what they want in a house. Let them. Take notes of what they say. If one person mentions hardwood floors, write it down. Pay close attention to what you're being told. As your sales skills improve, you will learn to read between the lines. It won't take long to determine which half of a couple has the most influence in the decision-making process. You must learn to assess customers quickly in order to appeal to them. While you might not become lifelong friends, your sale will go much smoother if you develop a friendly business relationship.

Don't be in a hurry to start talking about technical aspects of the home your customers want to build. Let time be on your side. Warm up to the customers. Seed the conversation with facts about you and your company that will build credibility and confidence. It's much more effective to weave this type of information into a friendly conversation than it is to run down a mental checklist, spewing out rehearsed lines one after another; avoid bragging about your accomplishments and just stick to the facts.

If your customers want to get right down to business, just do it. Be versatile. Not everyone is going to want to discuss their dogs, their children, or other aspects of their personal life. Some people will want to cut to the chase and discuss the specifications for their new home, so let them talk. You have to make your customers comfortable.

Your first goal during a sales meeting is to put your potential customers at ease. One way to break the ice is to give the customers books of house plans to look through; while they are skimming the pages, you can be talking in the background. Don't be answering phone calls during your meeting and avoid having your office staff interrupt you. Make those customers comfortable.

One of the most effective ways to keep people at ease is to meet with them in their home. People are always more at ease when they are in familiar surroundings. If customers come to your office, they may feel uneasy. Another advantage to meeting people in their home is that you can see how they live and what's important to them. Pictures on the walls will tell you a lot about your prospects. Even the furniture in the home will give you a hint about the personality of your customers. Use every advantage you can to gain a sale.

PRO POINTER

Most people like to talk about what they want in a house. Let them. Take notes of what they say. If one person mentions hardwood floors, write it down. Pay close attention to what you're being told. As your sales skills improve, you will learn to read between the lines. It won't take long to determine which half of a couple has the most influence in the decision-making process. You must learn to assess customers quickly in order to appeal to them.

Getting Down to Business

Getting down to business with customers seeking a custom home is not difficult. Start by asking them what type of house they like and the number of bedrooms and bathrooms that they would want. Don't talk about money, references, or workmanship in the beginning. Get the ball rolling with conversation that is pleasant for your customers. Let them turn on their dream machine, and then you can adjust the focus as the meeting goes on.

By the time your customers have finished describing their dream home, they should be excited and receptive to what you have to say; this is when you begin to go into details. If you have stock house plans, get out the ones that are closest to what the customer wants. Start talking about such features as quarry tile in the foyer and wallpaper and chair rail in the dining room. Play up the features of your homes.

When you are describing what makes your homes different and better, use photographs as visual aids. Pictures can do a lot to gain a person's trust. Even if you have never built a house, you can create a photo album that will help you sell your services. How can this be? I'll tell you.

Consumers tend to be better informed than they once were about subjects that interest them, and you as a builder need to be familiar with many aspects of the business. Many people are interested in the nuts and bolts of home construction, and you need to be prepared to discuss a wide range of technical topics. Gone are the days when you could give a broad-brush description of a home and escape without getting into the finer points. It helps to have lots of information at your fingertips. When dealing with first-time homebuyers who have limited funds to spend, you might be able to advise them of costs to add on to their home in the future (and create another potential sale in the next year or so). For example, adding a deck will increase the value of your house substantially; a family-room addition or a basement renovation will return 80 percent of your cost; adding a master-bedroom suite as the family grows will return nearly 80 percent. Information like this is readily available from the National Association of Home Builders and the National Association of Remodelers.

Know the product you are selling. I'm not talking about the nuts and bolts of construction—you already know that or you wouldn't be in the business—I'm talking about some of the economic advantages of owning a home. Particularly for those first-time homebuyers, you need to make them aware of the economic advantage of owning a home—the tax advantages of home ownership:

- Interest paid on mortgages is tax-deductible.
- Real-estate taxes are deductible.
- Points, the fees a bank charges a homeowner, in some cases are also deductible.
- Low-income buyers can contact their state or local government to see if they qualify for a tax credit that covers a portion of the mortgage interest.

PRO POINTER

When you are describing what makes your homes different and better, use photographs as visual aids. Pictures can do a lot to gain a person's trust. Even if you have never built a house, you can create a photo album that will help you sell your services.

- With each mortgage payment, equity is being added to the home; at some point in the future, home-equity loans may be available for home improvements.

- Interest on up to $100,000 of equity loans secured by the taxpayer's primary residence may be deductible.

- When the home is sold, the sellers can keep up to $250,000 of capital gains tax-free if they owned their home for two of the five years prior to its sale; the amount is double ($500,000) for married couples.

As you can see, there are a number of tax benefits accruing to the homebuyer, and you as a knowledgeable builder, by pointing them out to a prospective buyer, might be able to clinch the deal.

If you set up a photo album and what I call a prop box containing some samples of construction materials, you can impress customers very quickly. It's safe to assume that most builders that the customers talk to will not go to the same lengths to secure a sale. Here's how I do it.

PRO POINTER

If you set up a photo album and what I call a prop box containing some samples of construction materials, you can impress customers very quickly. It's safe to assume that most builders that the customers talk to will not go to the same lengths to secure a sale.

When I get to the point with customers of discussing details, I bring out the photo album. I show them pictures of houses that I've built, but I concentrate on the construction components. The closeup photos of pieces and parts capture the attention of most serious buyers. My photo album contains an assortment of closeup shots. For example, I show customers a picture of how my carpenters frame a header. You can build a small wall section with a header in your garage or lawn so it can be photographed. People will assume that it's part of a house you've built, whether it is or not.

Product suppliers and manufacturers are happy to provide builders with brochures and photos. Show your customers a detailed closeup of the cabinet construction that you can furnish in their new house. Point out the metal guides and glides that will make the drawers work smoothly. Concentrate on as many details of quality construction as time will allow.

While you have your customers looking at things, bring out your prop box. Hold up a metal joist hanger and explain what the item is, how it works, and why it makes for more solid construction. Next, pull out an electrical outlet and show the customers how your electrician uses only copper wire and how the wires are always installed under the screws for a better and safer connection. Show people the right and wrong way to install electrical outlets and switches, and explain how an improper installation can be dangerous, since the wires sometimes work loose if they are not under the screws. Don't talk down about your competitors, but let the customer know that you know how to do things the right way. Continue this type of show and tell until you sense that your customers are tiring of their education.

Have plenty of catalogs and brochures on hand for your customers to look at. Generating ideas can be an important part of a sales meeting when you are selling a house that doesn't yet exist. Looking at catalogs is a great way to maintain a level of enthusiasm in your customers.

Once you have convinced the customers that you are the best builder in the world, hand them an easy-to-read, bulleted checklist of the most important features incorporated in your homes. List the energy efficiency, the extended warranty (if you offer one, and you should), and any other special features. Never stop selling—but also know when to stop.

When you feel that you have your customer's confidence, move towards your closing pitch. Maybe you will give your customers a pencil and allow them to mark up a set of plans so that their modifications can be seen. If you have a CAD system on your computer, use it to help your customers get the layout they want. Put some paper in their hands. All you have to sell is your words and some paper, so make the most of what you have to work with.

If your meeting is going well, you should be getting close to the point of talking about money and financing. You may have to arrange a second meeting to get into financial details, but if you're selling stock plans with a price that you've already established or options that you've already priced, seize the moment and keep your momentum rolling. Your goal under these circumstances is to get a signature on a contract before you part company with your customers.

Is it really possible to sell an unbuilt house to people in just one meeting? It certainly is. I've done it on numerous occasions. Some people will want to talk with their parents or attorney before signing anything, and some people will want to think over your proposition. But a percentage of your customers will be ready to sign a contract right on the spot. Never be afraid to ask for a sale. You will never make many sales without asking people to buy from you.

The Key Elements

Let's go over the key elements involved with selling from blueprints. First, make your customers comfortable. Gain their confidence. Get them excited about the house they have been thinking about. Use visual aids to impress your customers. Remember to listen carefully to what your customers have to say, and by all means let them talk. Weave your sales pitch into a normal conversation. Don't use a canned pitch that makes you sound like a telemarketer. Stress your strong points and don't berate your competitors. It's okay to say that some builders do this or do that, but don't name names. It will make you look bad and perhaps desperate. Maintain control of your meeting at all times, but do it in a tactful way. Last, but far from least, ask the customers to make a deal with you before they leave.

I can't begin to remember how many houses I've sold from the hood of my truck and from various dining-room tables. It has been a lot of houses, I know that much. Usually, if I can keep customers in a sales meeting for at least two hours, I get a sale. Not always on the spot, but before the game is over I usually win. It often takes three or more meetings to hammer out enough details to close a sale, so don't be too impatient. On the other hand, never assume that you will have to go through an obligatory series of meetings to get a sale. Sell softly from the start and never stop selling yourself.

It will take time for you to perfect a personal style that is comfortable for you. Once you get the hang of selling houses "in the dirt," you'll be on your way to becoming one of the most successful builders in your area. If you just don't like selling, hire someone who can and will sell for you. Without sales you will have no business.

PRO POINTER

Let's go over the key elements involved with selling from blueprints. First, make your customers comfortable. Gain their confidence. Get them excited about the house they have been thinking about. Use visual aids to impress your customers. Remember to listen carefully to what your customers have to say, and by all means let them talk. Weave your sales pitch into a normal conversation. Don't use a canned pitch that makes you sound like a telemarketer. Stress your strong points and don't berate your competitors. It's okay to say that some builders do this or do that, but don't name names. It will make you look bad and perhaps desperate. Maintain control of your meeting at all times, but do it in a tactful way. Last, but far from least, ask the customers to make a deal with you before they leave.

Working with Real-Estate Brokers

Working with real-estate brokers is a subject that draws a lot of fire from builders. Some builders swear by brokers; others swear at them. A good broker can really help your building business, but an incompetent broker can break your bank account. Whether you are selling in the dirt or selling on spec, you have to perfect a plan that will work for your business. There's no clear-cut answer to what the best way to sell houses is. Each builder and every selling situation varies to some extent.

There are two basic types of real-estate agents; both work on a commission basis, collecting their fees when the sale is consummated and closing takes place. The first group represents the seller, in this case you, the builder; the second group represents the buyer. Real-estate agents, also referred to as "brokers," handle sales representing buyers or sellers but not both at the same time. Their fiduciary responsibility is toward their client, either the buyer or the seller, and this is an important point to remember. There are several subgroups of real-estate professionals: Some specialize in multi-family housing; others concentrate on commercial properties. A few are generalists: people who will sell all types of real estate on commission. If you look hard enough, you might find a new-construction specialist. This is easier to do in major residential market areas but can be difficult in rural areas or areas where new construction is not in abundance.

As a builder, you have many options open to you when you decide to sell your houses. If you have the time, talent, and the desire, you can make the sales yourself. Some builders set up in-house salespeople to sell their houses, while other builders hire brokers. Which of these three ways appeals to you?

I've been in the building business long enough to experiment with all three methods of selling. My personal sales efforts have been very effective, but when I was building in high volume, I simply didn't have the time to make all the sales myself. Using real-estate agents to sell for me has worked with limited success, but in my case it was the least effective method. When I developed an in-house sales team, I saw good results when I was building in volume. To define your needs, let's look at the three options more closely.

PRO POINTER

There are two basic types of real-estate agents; both work on a commission basis, collecting their fees when the sale is consummated and closing takes place. The first group represents the seller, in this case you, the builder; the second group represents the buyer.

Sell It Yourself

When you build a house and can sell it yourself, you will have more control over the sale and will avoid having to pay the agent a commission. This works for builders who are accomplished in sales skills and who are building a low volume of houses so they can maintain an active presence as both a builder and a sales professional. Unfortunately, many builders don't possess good sales skills. These skills can be learned, and I think every builder should develop at least a minimum sales ability.

One of the biggest advantages of selling your own homes is that you know what customers are being promised. I've had real-estate agents promise customers the moon to get a sale. This makes it very difficult for the builder, who has to live up to the expectations of the customers. If you're doing the selling, at least you know if you have to offer something more in order to make the sale.

Assuming that you either already know how to sell or are willing to learn, I think a start-up builder should try selling; it will put you in touch with buyers and shoppers and give you more insight into what potential buyers are looking for in a home.

Builders share different opinions on the best way to make a sale. I can tell you about my personal experiences and I can share recollections of other builders, but I can't tell you what will work best for you. You must assess all the data available to you when making a decision. If you feel uncomfortable selling, don't try to sell your own homes.

In-House Sales

An in-house sales staff is difficult for most new builders. The cost of the salespeople and the volume of sales that you hope they will create can cause a start-up builder more trouble than they're worth. Once your building operation is running smoothly, an in-house salesperson can be very valuable. But most builders have to grow into a position where having a salesperson on the payroll will make sense.

In-house sales associates are sometimes paid an hourly rate, just like a carpenter, but this is rarely the best way to structure a compensation package. Many big-league builders have their personal sales staff on a commission-only basis. If the salespeople don't sell, they don't get paid. This type of arrangement is ideal, but it's not practical for a small-time builder.

Commission-only salespeople need to sell a lot of houses to make a good living. They also need the houses completed and closed quickly so that they can generate income. The sales professionals usually don't get paid until a house is built and the property transfer is complete. Small builders usually don't have the ability to offer enough volume or quick enough turnaround time to attract top-notch, commission-only sales associates.

This leaves the small builder with the option of paying a draw against commission. The sales staff is technically paid by commission, but they receive regular paychecks to earn some income until house deals close. This requires a builder to incur the expense of the paychecks until deals close, and if you are working with a tight budget, you probably can't afford to put out money for salespeople before you have closed on houses. Herein lies the problem.

Real-Estate Agents

If you can't sell houses very well yourself and you can't find or keep experienced salespeople on staff, what's left? Independent agents and brokers are the remaining solution. Going to a broker offers a builder several advantages. One is the fact that representing your houses won't cost you a cent until a house is built, sold, and closed. There are, however, some disadvantages to using

> **PRO POINTER**
>
> If you can't sell houses very well yourself and you can't find or keep experienced salespeople on staff, what's left? Independent agents and brokers are the remaining solution. Going to a broker offers a builder several advantages.

agents, so let's get into the pros and cons of listing your homes with a real-estate broker.

The terms real-estate "agent" and real-estate "broker" are often used interchangeably, and from a practical standpoint they perform much the same function; however, the state licensing tests for agents are generally less complex than those required for brokers. And the broker in some states must have had some previous experience in real-estate sales working for a firm selling real estate. The term "Realtor®" designates a member of the National Association of Realtors.

PRO POINTER

The terms real-estate "agent" and real-estate "broker" are often used interchangeably, and from a practical standpoint they perform much the same function; however, the state licensing tests for agents are generally less complex than those required for brokers. And the broker in some states must have had some previous experience in real-estate sales working for a firm selling real estate. The term "Realtor®" designates a member of the National Association of Realtors.

Buyer's Agents

Buyer's agents, as you would expect, represent buyers rather than sellers; you as a builder would not use such an agent, but you could very well do business with him or her. If an agent has a client who wants to have a new house built, she might contact you for the purpose of providing estimates and proposals for your services. Keep in mind that the agent or broker is working with you but on behalf of the buyer.

Buyer's brokers can be paid by either sellers or buyers. This is a point that confuses a lot of people. Many people assume that since buyer's brokers work for buyers, they are paid by buyers. This is not always the case. If the broker's client wants to do business with you, the contract offer to purchase may stipulate that as part of the terms it will be your responsibility to pay the broker. This is not necessarily bad, but it is something that you should be aware of and understand.

Most builders include the cost of a real-estate commission in the sales prices of their houses. If a brokerage is used, the fee is paid. Builders who sell homes themselves pocket the extra profit. In my opinion, paying a reasonable commission to a buyer's broker for bringing a deal to the table makes a lot of sense. It's a sale you wouldn't otherwise have. For the sake of our conversation, it's enough that you understand that you may be asked to pay the fee of a buyer's broker, so look for the fine print in any offer that is presented to you by a broker or buyer.

Exercise of Option

To _____, Optionor, as Optionee, it is my intent to
exercise the option agreement dated _____, between the
Purchasers, _____, and the sellers
_____, for the sale of the real estate
commonly known as _____.
The escrow agent for this transaction is:
Name

_____

Address

_____

Phone

_____

The amount deposited in escrow is
_____ ($_____) and this deposit
was made on _____.

_____   _____
Optionor   Date   Optionee   Date

_____   _____
Optionor   Date   Optionee   Date

FIGURE 8-1 Sample of an option form.

Seller's Agents

Seller's brokers are the most common type of real-estate salespeople. These are the traditional brokers who work for sellers and with buyers. If you've been a builder for a while, seller's brokers asking you to list your homes with their firm have probably contacted you.

Is it a good idea to list your houses with a broker? It can be, but it can also put you in a difficult spot. If you list a house with a broker who is experienced and aggressive, you might see a contract in just a few days. But if the broker is inexperienced or not motivated to sell a house that is under construction, you could have your house on the market for months with no sales activity. Selecting a listing agent is tough but must be done right if you want to avoid all the headaches.

Some real-estate salespeople sell by mis-stating the facts in order to make a sale They don't lie to customers, but they may stretch the truth a little or make promises without contacting the builder first. They may make things sound a little better than they really are. A broker who operates this way can make your job extremely unpleasant. You may find that customers are coming to you complaining that their house wasn't built the way it was supposed to be. It will be your job to satisfy the customer or live with the bad reputation they are likely to give you. I've gone through this situation, and I know how miserable it can be. In fact, let me give you a few examples of what some of my experiences have been with independent brokers.

I remember one house where the broker who sold it told the customers that a dishwasher was included in the price. The spec sheet on the house included a dishwasher, but only as an add-on option. When this particular house was completed, a walk-through was done against the contractual specifications, and the customer complained to one of my field supervisors about the lack of a dishwasher. After checking into the matter, we learned what had happened, and rather than leave a customer with a bad taste for my company, I paid to have a dishwasher installed—and I never listed another house with that broker. I admit, a dishwasher was a fair trade for a good contract, but I should have been consulted before the deal was made.

I had another broker promise customers that my workers would help them unload and install their old washer and dryer in the new house. All of a sudden my field supervisors were turned into movers. Go figure.

My list of stories could go on, but I can tell you that some salespeople can create a lot of trouble for you when they start giving customers whatever it takes to make a deal without checking with you first.

Listing your homes with a seller's broker can be a good experience, but the key is in finding the right agent or broker and the best brokerage. To do this, you must understand a little bit about how the real-estate game is played.

PRO POINTER

Listing your homes with a seller's broker can be a good experience, but the key is in finding the right agent or broker and the best brokerage. To do this, you must understand a little bit about how the real-estate game is played.

Big Real-Estate Companies

Big real-estate companies have their advantages. National franchises get a lot of referrals, and they often produce a lot of prospects. You may think that you are better off to list your house with these big brokers than with smaller, local companies. This is not always the case. In fact, sometimes the opposite is true.

There is no question that being a household name is an advantage in the real-estate business. Several franchise names come to mind when I think about recent television commercials. If I were making a long-distance move, I might very well contact one of the national brokerage firms to help find my new home, but the big-name companies don't always do better for builders than a small company.

PRO POINTER

When you list a house with the large real-estate firms, your home might be only one listing out of 300 and might get less personal attention and exposure than it would at a smaller company. Listing your house with a one-person real-estate company could get you a lot more attention and advertising. It only takes one broker to sell your house. You don't need to list with an army of agents to get results. What you have to do is find the right combination of agent, broker, and company.

When you list a house with the large real-estate firms, your home might be only one listing out of 300 and might get less personal attention and exposure than it would at a smaller company. Listing your house with a one-person real-estate company could get you a lot more attention and advertising. It only takes one broker to sell your house. You don't need to list with an army of agents to get results. What you have to do is find the right combination of agent, broker, and company.

Ask Questions

Before you agree to list your house with a broker, ask a lot of questions. How many new-construction homes has the firm sold in the last 90 days? If the firm has not sold many, it is probably the wrong one for you. Selling from blueprints and shell homes takes a different type of talent than what is required to sell a completed home. You need to list with people who have what it takes to sell what you have to offer. There are a number of brokers and brokerages to choose from, so keep searching until you find the right one.

Broker Commission Arrangement

If _____ , broker of the

_____ agency procures

an acceptable offer for the purchase of my real estate, commonly known as

_____ , and the property

is successfully sold, the real estate agency shall receive a commission equal to

_____% of the closed sale price. The listed price of this property is

_____

($_____). This commission agreement will remain in effect from

_____ to _____. Seller agrees

that if the property is sold within six months to anyone the broker has registered

with the seller, as a prospective buyer, the broker shall be entitled to the above

commission. This does not apply if the seller lists the property with a licensed real

estate brokerage on an exclusive basis.

_____   _____
Seller   Date   Broker   Date

_____
Seller   Date

FIGURE 8-2   Sample of a commission agreement to use with real-estate brokers.

Other questions to ask are: How often will your home be advertised? Will it be a spotlighted listing? What media will be used to promote your home? Ask these questions and get the answers. If the answers are not what you want to hear, keep looking for another broker.

What type of listing are you being asked to sign? An open listing allows anyone to sell your house. This is good in one way, bad in another. The good thing is that you are free to sell the home yourself without paying a commission, but because most brokers will not advertise open listings and can lose the commission to another agency, they may not be motivated to seek a sale.

An agency listing is, in my opinion, the best option for a builder. This type of listing gives a broker an exclusive listing agreement that protects the firm. Only the listing agent can take credit for a sale, with the exception of you or your in-house sales staff. If you sell the house, you don't have to pay a commission. If another agency brings a buyer in, the listing agency gets a split of the commission, usually half. This type of listing provides some protection for the brokerage and leaves you free to sell your own property without paying a fee.

An exclusive listing is what most brokers offer. The broker will get a commission regardless of who sells the house, even if you sell it yourself. The good thing about an exclusive listing is that the brokerage will work harder to make a sale, but this doesn't always happen. If I were you, I'd stick with agency listings, so that I could sell my own houses without paying a fee.

Regardless of what type of listing you choose, make sure you understand all the terms and conditions. Read the listing completely, even the boilerplate, and when you are finished, if you have any questions, don't hesitate to ask for an explanation. For example, how long will the listing run for? Most of them go for four to six months. Some are shorter and some are longer. Can you terminate the listing at any time if you are not pleased with the service you are getting? Will you have to pay a termination fee if you stop the listing early? Ask these questions. Get a listing that you can terminate at any time without paying a penalty fee.

The Commission

The commission you agree to pay a brokerage is strictly up to you and the brokerage. The amount could be 3 percent or 10 percent. New homes often pay a broker 5 percent, but 6 or 7 percent is not uncommon. Higher commissions

PRO POINTER

The commission you agree to pay a brokerage is strictly up to you and the brokerage. The amount could be 3 percent or 10 percent. New homes often pay a broker 5 percent, but 6 or 7 percent is not uncommon. Higher commissions are rare, and so are lower ones. There is often no set rule as to what a commission has to amount to, but local real-estate associations set fees for all their member firms, and anyone deviating from this fee structure is asked to leave the association. So ask about fees; they may be negotiable.

are rare, and so are lower ones. There is often no set rule as to what a commission has to amount to, but local real-estate associations set fees for all their member firms, and anyone deviating from this fee structure is asked to leave the association. So ask about fees; they may be negotiable.

When you are checking out a brokerage, see how many of their in-house listings are sold by their own people and how long, on average, their listings have been on the market. A broker who sells his own listings should be a good one to go with.

PRO POINTER

When you are checking out a brokerage, see how many of their in-house listings are sold by their own people and how long, on average, their listings have been on the market. A broker who sells his own listings should be a good one to go with.

Multiple Listing Service (MLS)

Most real-estate brokerages belong to some type of multiple-listing service (MLS). Joining a MLS isn't mandatory, and a broker can sell effectively without the benefit of MLS services, but I believe that you will be better off if you list with a brokerage that uses an MLS system. When a broker lists your house in a MLS system, other real-estate firms receive the listing. This can reduce the house's time on the market. You want quick sales, so look for all the advantages that you can find.

Doing the Grunt Work

Brokers do the grunt work for builders. It is the broker who pays to advertise a house. When phone calls come in, the broker handles them. Going out in the evening to show houses is a responsibility that good brokers assume. You, as the builder, just wait for the broker to bring you a purchase offer. A good broker should prequalify prospects, help

PRO POINTER

Most real-estate brokerages belong to some type of multiple-listing service (MLS). Joining a MLS isn't mandatory, and a broker can sell effectively without the benefit of MLS services, but I believe that you will be better off if you list with a brokerage that uses an MLS system. When a broker lists your house in a MLS system, other real-estate firms receive the listing.

with financing, and serve as a buffer between you and the customer. When the system works well, brokers are a favorable asset, but when the system doesn't work, don't judge the entire industry by the actions of one agent.

Most builders, especially new builders who aren't comfortable selling, should list their homes with brokers. If you do, there may be some problems, but the good should outweigh the bad if you do enough homework to select an ideal broker.

PRO POINTER

Most builders, especially new builders who aren't comfortable selling, should list their homes with brokers. If you do, there may be some problems, but the good should outweigh the bad if you do enough homework to select an ideal broker.

Finding and Selecting Green Building Products

When you venture into sustainable building, you will need green building products to incorporate into your projects. What good would a green builder be without green products? If you are new to the building business, this may not seem like much of a challenge to you. However, if you are an old dinosaur like I am, the transition can be a bit more trying. Don't despair. Finding and selecting green building products need not be a nightmare. In fact, it can be quite easy. What are your options? They are numerous. Let me list a few of them for your fast consideration:

- Online Internet searches
- Printed directories
- Organizations
- Online forums
- Online newsgroups
- Manufacturers
- Suppliers
- Integrated systems

These are just some of the many ways to outfit yourself with suitable sustainable materials.

Using a Computer

Using a computer to find what you are looking for is probably the fastest, most efficient way to acquaint yourself with the new world of green building materials. There is little doubt that an online search is a key way to get the job done. However, there are those of us who still prefer paper over pixels.

Older builders are sometimes resistant to using a computer for anything. At best, they have people for that. For some builders this is an option, but if you don't keep up with current events you are often left out of conversations. The same is true when it comes to using a computer in the building business. Just because you have built homes for 20 years without relying on computerized data doesn't mean that you should go forward in the green building world without a high-speed Internet connection.

I still remember the first computer I used in my business. Believe it or not, it was a Commodore 64. Yeah, I am an older guy. When my wife pushed for the computer, I fought it. Fortunately, she won, and I became converted. Looking back to those past decades, I can't be too thankful that I learned to use a computer before it was too late. Even so, there are those who simply will not plug into the online community.

Assuming that you are computer-literate, you know about search strings and search engines. These services can tie you into green products and suppliers in moments. Obviously, this is the quickest way to get a jump-start in your education on green products.

Typing a few words into a search engine will generate results. What will you do with them after you obtain them? Knowing that products exist and knowing the products are two very different issues. Let's look at an example of what I am talking about.

One of my last projects involved the installation of bamboo flooring. I worked with my regular flooring supplier/installer to accomplish the job. But, supposing that I was new to sustainable products, what would I have done? I would have done my due diligence in research and then reached out for expert assistance. How would I go about the research? Personally, I would consult both Internet and printed sources, as well as local suppliers. I prefer to have multiple options before casting my vote.

FIGURE 9-1   Flooring offers excellent options for going green (courtesy of Natural Cork and More).

Assume that you don't know anything about environmentally friendly building products. You have a customer who is interested in bamboo flooring, and you have a meeting with the customer next week. How are you going to get up to speed on current trends? If you use the Internet, you can get some strong talking points. For this example, I did a quick search and came up with plenty of information to help me sound as if I know what I am talking about even if I don't. Let me list some of the talking points I found that I could use in a meeting with a customer.

- Installation areas must be maintained at normal occupancy temperature and humidity levels for at least 10 days.
- Flooring material must be maintained at normal occupancy temperature and humidity levels for at least 10 days.
- Any adhesives used in the installation must be maintained at normal occupancy temperature and humidity levels for at least 10 days.
- Room temperatures are most suitable when they range from 60–70 degrees Fahrenheit.
- Relative humidity should have an ideal range of between 40–60 percent.
- The maximum range of temperatures in a room should be between 50–80 degrees Fahrenheit.
- The maximum range of humidity is required to remain between 35–65 percent.
- A humidifier or dehumidifier can be used to maintain a suitable range of humidity.
- Bamboo flooring, when cared for within manufacturer's recommendations, has an expected life of 20 years.
- Color variations may exist in the flooring material.
- The edge of the flooring is tongue-and-groove with a bevel edge.

I could go on and on, but you get the idea. There was plenty of additional information readily available to me. I am sure that in less than one hour I could be prepared to talk a good game with a new customer. Talking the talk, however, is not the same as walking the walk. An occasional bluff may be needed, but you have to have the knowledge to back up the bluffs. In short, don't con your customers. You can bet that many of them have done as much homework, or more, than you have.

Putting together a pitch book to make sales is easy. Turning the sale into a successful building project will require much more education. Fortunately, the education is essentially free of charge and available online. Dedicate some of your time to learning about sustainable building from a variety of sources. Retain what you learn.

FIGURE 9-2 Research of green products prior to installation is key to success in green building (courtesy of Superior Walls of America, Ltd.).

Creating Your Research Library

One excellent way of becoming an expert in your field is by creating your own research library. This can be done on a computer or on paper. I prefer paper, but others prefer electronic data. Saving a tree is worthwhile for sure, but I like to be able to go to a shelf and put my hands on information. How you find, store, and review your research materials is a personal choice. The key is doing it.

Do your online research and save the data obtained. I often print out the research pages and save them in an appropriate binder for future use. Some builders save the files electronically, which is fine. With my personal system, the research is grouped by category and stored for future reference. If a customer wants to know about roofing materials, I have a binder for it. When the topic is low-water-usage plumbing fixtures, I have a binder for it. Not only do I have it; I have studied the material and absorbed much of it. This enables me to be knowledgeable when discussing sustainable building practices and materials.

PRO POINTER

Product suppliers are happy to provide free information on the products that they market. Check with your local suppliers. After all, they are most likely the people you will be buying from. See what is selling best. They will be happy to tell you. Ask for brochures, installation instructions, Web sites, and so forth. All of this information can go into your personal library.

Product suppliers are happy to provide free information on the products that they market. Check with your local suppliers. After all, they are most likely the people you will be buying from. See what is selling best. They will be happy to tell you. Ask for brochures, installation instructions, Web sites, and so forth. All of this information can go into your personal library.

Books

Much of what I know has been learned from books, and I continue to turn to them for knowledge. Some say books will go the way of the dinosaurs (like me), but I believe there will always be a demand for tangible words that can be touched, pages that can be turned down, creases to hold scraps of paper as bookmarks, and door stops. Books can do all of this and more.

There is no shortage of books on the booming green building business. A trip to the bookstore or a cruise around the online bookstores will show you plenty of choices for buying books on sustainable subjects. Books speak for themselves. You can learn from the experience, failure, investments, and rewards of experts for a very small price. In short, books are a tremendous value when you compare the cost versus the return. My favorite source of this type is the seventh edition of the Greenspec Directory. It is published by BuildingGreen, Inc., and offers a full array of information for the green industry. The book is not cheap, but it is a good value.

Organizations

Organizations are usually a good source of dependable information. The green industry is no exception. An online search will reveal numerous organizations for sustainable builders. Traditional organizations, such as the National Association of Home Builders, are still strong resources for new building trends.

Look into local organizations and see what you can participate in. If you live in a small community that does not have a strong base of building groups, consider joining long-distance organizations. Most groups produce routine mailings and updates to keep you abreast of changes in the industry. The dues are sometimes a deterrent, but the benefits often outweigh the cost.

PRO POINTER

Organizations are usually a good source of dependable information. The green industry is no exception. An online search will reveal numerous organizations for sustainable builders. Traditional organizations, such as the National Association of Home Builders, are still strong resources for new building trends.

Suppliers

Suppliers are your partners in sustainable building practices. Not all suppliers have joined the green movement, but most have. As a green builder, you need green suppliers. Customers will frequently want to see what they are buying. Suppliers who offer showrooms can be of great help with it comes to touching what has yet to be built. Get out and survey suppliers of green building materials. Determine which ones you will work with.

Once you know which suppliers fit your needs, it is time to get personal. It will take some time and time is money, but so are relationships when it comes to customers. Meet the representatives of the suppliers. Go to lunch with them if you have to. Spend time with the salespeople. Let them get to know you. Build a solid relationship that will show when your potential customers go to the supplier to look for options on building materials.

When you think of suppliers, do not limit yourself to lumberyards. Some carpenters who become builders fall into this trap. Go the distance. Take the time to interview suppliers of all the materials that will be used in your projects. This includes plumbing, heating, and electrical suppliers as well as interior elements such as lighting fixtures.

Subcontractors

Nearly all builders use subcontractors for some portion of work being done on a new project. Just because you have gone green doesn't mean that your subcontractors have.

FIGURE 9-3 Water-saving plumbing fixtures should be incorporated into sustainable building (courtesy of Zurn Industries, LLC).

FIGURE 9-4   Sensor-operated plumbing fixtures can reduce water consumption and waste (courtesy of Zurn Industries, LLC).

FIGURE 9-5 New technology in toilets allows for maximum utilization of water resources (courtesy of Zurn Industries, LLC).

This can be a problem when it comes to customer satisfaction. It is hard enough to find reputable, dependable subcontractors, but finding them with green credentials can be even more difficult. As the builder, it is your responsibility to provide your customers with suitable subs. Get it right, or you will regret it.

Call in subcontractors for meetings after you have a firm grasp of green building. Interview the subs. See what they know and who they work with. Will they enhance your image or hurt it? Don't associate your company with the wrong subcontractors; it can take you down.

Integrated Systems

If you want the easy way out, consider integrated systems. These types of systems are fairly new, but they are becoming popular. I know a builder who has developed a plan for traditional builders to turn green almost overnight. Essentially, this builder has created a system that he has perfected to a point at which almost any builder can be a responsible, sustainable builder by following a proven method. This minimizes the research and effort on your part. The cost of admission for an integrated program can be steep, but the rewards can be great. It is worth considering.

PRO POINTER

If you want the easy way out, consider integrated systems. These types of systems are fairly new, but they are becoming popular. I know a builder who has developed a plan for traditional builders to turn green almost overnight. Essentially, this builder has created a system that he has perfected to a point at which almost any builder can be a responsible, sustainable builder by following a proven method. This minimizes the research and effort on your part.

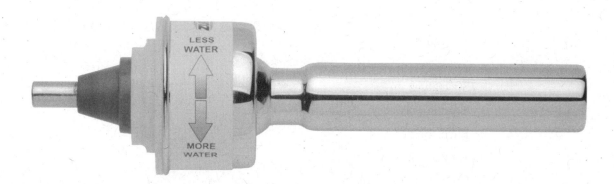

FIGURE 9-6   In-line controls allow excellent control over water flow in buildings (courtesy of Zurn Industries, LLC).

Types of Materials and Systems

There are varied types of materials for sustainable building. Some to consider are listed below:

- Waste management
- Air quality
- Green building materials
- Used building materials
- Certified wood materials
- Energy conservation
- Renewable energy equipment
- Recycling programs

To give you an idea of the types of materials that you could be dealing with as a green builder, consider the following list:

- Concrete
- Masonry
- Metals
- Wood
- Plastics
- Composites
- Thermal protection
- Moisture protection
- Openings
- Finishes
- Specialties
- Equipment
- Furnishings
- Special construction
- Conveying equipment
- Plumbing
- HVAC

- Electrical
- Communications
- Security
- Earthwork
- Exterior work
- Utilities
- Transportation
- Pollution control

This list is far from conclusive. When you embark on a career as a green builder, you have some research to do. Thinking like a builder, you should consider the following:

- Sitework
- Landscaping
- Decking
- Outdoor work
- Foundations
- Footings
- Slabs
- Structural components
- Sheathing
- Exterior finish
- Exterior trim
- Roofing
- Windows
- Doors
- Insulation
- Flooring
- Floor covering
- Interior finish
- Interior trim

- Paint
- Coatings
- Caulking
- Adhesives
- HVAC
- Plumbing
- Lighting
- Electrical
- Appliances
- Furniture
- Furnishings
- Energy usage

There you have it. As you can see, there is much to learn in the green industry. The sooner you learn it, the sooner you can cash in on it. Not only will you be likely to make more money, you will be ensuring a better environment for generations to come.

Dealing with Subcontractors and Suppliers

Subcontractors and suppliers are major players in the contracting businesses. When contractors know how to work with and control these participants in their business, their businesses prosper. Hiring subcontractors who perform shoddy work will reflect directly on how the public views your company-as one that does not care about the quality of its product. When suppliers repeatedly fail to deliver quality products on time, your crews can come to an abrupt halt. If you want your business to be successful, you must develop strong relationships and learn how to work with subcontractors and suppliers. In this chapter we will deal with each of these aspects of your business; let's start with subcontractors, who will share a large role in your day-to-day business activities.

Subcontractors

As a contractor, it is common practice to hire subcontractors to perform various forms of work, typically site work, underground utilities, plumbing, HVAC, electrical, and often roofing and flooring. The reputation for quality of these subcontractors is very important, since their work will have a direct reflection on your company. The average homebuyer does not know who installed their plumbing work or who tiled the bathroom. They only know that it looks good, a reflection on you, or it looks substandard, also a reflection on you, the builder.

Subcontractors frequently have contact with your customers, and you should expect them to maintain the reputation you've worked hard to earn. However, without care in selecting a subcontractor and checking his work from time to time, you can expect some problems down the road. When you have a core group of dependable subs, you can respond to work quickly and efficiently. Customers love to get fast service, and subcontractors can give you this desirable dimension.

Subcontractors with good work habits and people skills will build your business. A lot of subs don't want all the responsibilities that go along with being the general contractor. If you select good, reputable subcontractors and treat them as you would your own employees, they will help you prosper as they prosper.

> **PRO POINTER**
>
> Subcontractors frequently have contact with your customers, and you should expect them to maintain the reputation you've worked hard to earn. However, without care in selecting a subcontractor and checking his work from time to time, you can expect some problems down the road. When you have a core group of dependable subs, you can respond to work quickly and efficiently.

Suppliers and Vendors

You might not think that suppliers (or vendors, as they are often called) can have a major impact on your reputation, but they surely can. As a contractor, you are responsible for everything that happens on the job. If your supplier's delivery truck damages the customer's lawn, you're going to hear about it quickly, and the customer will look to you to repair the damage, even though you will expect the supplier to reimburse you for these expenses. When materials are not delivered on time, customers are not going to call the suppliers to complain; they are going to call you. As the general contractor, you take all the responsibilities and all the complaints.

If you want your customers to remain happy—and what contractor doesn't?—you must be in control of all aspects of the job, from getting the permit to doing the punch-out work and everything in between.

When you have reliable suppliers, they can improve your customer relations. When delivery drivers are courteous and professional, customers will appreciate it. When deliveries are made on time, customers are satisfied. Seeing to it that suppliers make and maintain good customer relations is up to you. You have to establish the rules for your suppliers to follow. If the suppliers are unwilling to abide by your rules, find new suppliers. Remember, in this case, you are the customer!

Materials

Delivery of the correct materials, in good shape and on time, can have a positive impact on your customer's peace of mind. If framing lumber, for example, shows up bowed with lots of splits or knotholes, your customer isn't going to be pleased. If the wrong materials are delivered and your crews are unable to continue working, you will lose time, and your customer will lose patience. Don't overlook the important role that materials play in the way customers view your business.

Choosing Your Product Lines

Choosing your product lines carefully is important to the success of your company. If you pick the wrong products, your business will fizzle. When you choose the proper products, ones that are familiar to your customers, they will sell themselves. As a business owner, you can use all the help you can get, so carry products that the public wants.

> **PRO POINTER**
>
> Choosing your product lines carefully is important to the success of your company. If you pick the wrong products, your business will fizzle. When you choose the proper products, ones that are familiar to your customers, they will sell themselves. As a business owner, you can use all the help you can get, so carry products that the public wants.

You can select products by doing some homework. Read magazines that appeal to the type of people you want as customers. For example, if you want to become known for the outstanding kitchen and bathroom designs used in your homes, read magazines that focus on kitchens and baths. Observe advertisements in the magazines. By paying attention to these ads, you will get a good idea of what your customers will be interested in.

Walk through the local stores that carry products you will be using or competing with. Take notes as you walk the aisles. Pay attention to what is on display and its cost. Take note of products that you see on television or advertised in the paper or in magazines. Your potential customers probably watch those same programs, read the same local paper, or subscribe to some of those magazines. This type of research will help you to target your product lines.

Visit some of your competitors' projects, particularly the successful ones. Go into their sample homes and check out the appliances that are installed, the type of flooring, the paint colors, the windows and doors. Look at homes under construction. See what your competition is doing. By simply

PRO POINTER

Visit some of your competitors' projects, particularly the successful ones. Go into their sample homes and check out the appliances that are installed, the type of flooring, the paint colors, the windows and doors. Look at homes under construction. See what your competition is doing.

riding past a construction site you can probably tell what types of doors, windows, siding, shingles, and similar items are being used. This on-site investigating can put you in touch with what the public wants.

When you talk to your customers, ask them about products that they'd like to see in their home. By doing so you not only get some valuable information but you also give your customers the feeling that you really care about being up to date on current market trends.

There are several ways in which you can conduct informal surveys. You can go door to door and do a cold-call canvassing of a neighborhood. You will experience a lot of rejection, but you will also get some answers. If you don't like knocking on doors, you can use a telephone. You can even have a computer make the phone calls and ask the questions for you. You can also use direct mail. Direct mail is easy to target, and it's fast and effective. While mailing costs can be expensive, the results are often worth the expense. You could design a questionnaire to mail to potential customers. Done properly, your mailing will look like you have a sincere interest in what individuals want. It will appear this way because you will have a sincere interest.

Answers to your questionnaire will tell you what products to carry. You can improve the odds of having the pieces returned by self-addressing the response card. You can absorb the cost of the return postage by purchasing a permit from the local post office and having it printed on your cards. You can affix postage stamps to the cards, but this will cost more. With the permit from the post office, you pay only for the cards that are returned, not counting the permit fee. If you use postage stamps, you will pay for postage that may never be used.

To convince people to fill out your questionnaire, you need an incentive such as offering a discount from your normal fees. A better idea might be to make the questionnaire look more like a research project, which in fact it really is. If you design the piece to look like a respectable research effort, more people will respond to your questions and feel that you are a progressive businessperson interested in learning what the public really wants. Many new sustainable products are becoming available quickly. What was new a few months ago might have significant competition by now. Do your homework. Join some green organizations and stay on top of what is new in the business. Choosing to go green means that you have to become an authority on the products available to you and your customers.

Today's buyers are knowledgeable. You can bet that most of them have done their research. You will be embarrassed if you discover that a potential customer is more astute than you when it comes to environmentally friendly options. I can't stress this enough. Eat, sleep, and breathe green until you know it, and then stay on top of it. Set aside a couple of hours a week for research. With the bursting growth of the industry, you can't settle in and be done with it.

Avoiding Delays in Material Deliveries

Avoiding delays in material deliveries is crucial to the success of your business. If your materials are delayed, your jobs will be delayed. If your jobs are delayed, your payments will be delayed. If your payments are delayed, your cash flow and credit history can falter, and if your cash flow slows down or disappears, your business is headed for trouble. How do you avoid late material shipments? You can maintain acceptable delivery schedules by being involved personally.

Your job of maintaining the delivery schedule begins when you place the order. Some basic principles apply to keep your deliveries on schedule. First of all, place your orders far in advance of when you need them, thereby giving your supplier sufficient time to work it into the delivery schedule. Get the name of the person taking your order. Use a phone log to

PRO POINTER

Avoiding delays in material deliveries is crucial to the success of your business. If your materials are delayed, your jobs will be delayed. If your jobs are delayed, your payments will be delayed. If your payments are delayed, your cash flow and credit history can falter, and if your cash flow slows down or disappears, your business is headed for trouble.

document all your calls. Ask the order taker to give written documentation of your delivery date. While you are at it, get the name of the store manager; you may need it if you have problems with the delivery. When you have the intended delivery date, stay on top of the delivery. If you have placed the order several days in advance, make follow-up phone calls to check the status of your materials. Always get the names of the people you are talking to; you never know when you will have to lodge a complaint. Keep clear records of your dealings in an order log. By maintaining a telephone or personal presence you will establish a relationship with that supplier and the employees handling your deliveries. They will assume that if you are this attentive now, you will be tough to deal with if the order gets screwed up.

With a lot of effort and a little luck, your deliveries will be made on schedule. If the shipment does go astray, contact the store's manager. Advise the manager of the problem and the ripple effect it is creating for your business. Produce your documentation on the order. By showing the manager your written delivery date, employee names, and supporting documentation, such as material specifications, you will make a strong impression. This tactic will set you apart from the customers who complain but can't back up their complaint with facts. You will come across as a serious professional. If you don't get satisfactory results, move up the ladder to higher management or consider finding another source and let your current supplier know that you are looking around.

Choosing Subcontractors

Choosing subcontractors requires lots of time and effort. This is a key element in your business and requires a lot of consideration. Your business reputation will rise or fall based upon the quality of your subcontractors.

> **PRO POINTER**
>
> Choosing subcontractors requires lots of time and effort. This is a key element in your business and requires a lot of consideration. Your business reputation will rise or fall based upon the quality of your subcontractors.

The Initial Contact

You will learn a lot about your subcontractors from the initial contact. Did they show up on time and have a neat physical appearance? If so, they're off to a good start. Your first impression may not be accurate, but you are sure to formulate one.

Not only will you form a first impression of the subcontractors, but they will also form an opinion of you, so you should plan to be free when they arrive at the appointed time. Make sure that your office is neat and has an organized appearance. The potential importance of this meeting suggests that you should take control of the proceedings.

As a contractor, you will come to rely on subcontractors, and you need to make a good impression. Have the topics you wish to discuss at hand and discuss them in a professional manner. Just as subcontractors may alienate you, you may alienate them. This, obviously, is not what you want to do. You want your first contact with subcontractors to be productive so you need to choreograph it carefully. Wasted time is potentially wasted money.

Just as you want to engage professional subcontractors, subcontractors want to work with successful contractors. One of subcontractors' greatest concerns is whether or not they will be paid promptly or even if they will be paid in full. If, as a general contractor, you come across as an unorganized, financially shaky business, subs will not be thrilled at the possibility of working with you.

PRO POINTER

Subcontractors and general contractors go together like peanut butter and jelly. If there is not a comfort level between the two parties, the working relationship will not succeed. Plain talk and honesty are the best traits to exhibit when talking to subcontractors, and you should expect the same from them.

Subcontractors and general contractors go together like peanut butter and jelly. If there is not a comfort level between the two parties, the working relationship will not succeed. Plain talk and honesty are the best traits to exhibit when talking to subcontractors, and you should expect the same from them.

What can you do to attract quality subcontractors? If you project a professional image, one with a sound business philosophy, subcontractors will seek you out. Whether you are talking on the phone or in person, send the right messages. Let subcontractors know you are a professional and will accept nothing less from them.

Application Forms

Application forms can come in handy when interviewing subcontractors. While subs are not going to be traditional employees, it is not unreasonable to ask them to complete an employment application. The applications used may not resemble those used for employees, but you want to know as much about them as possible.

The application should contain questions pertaining to the type of work the subcontractor is equipped to perform. Asking for credit and work references is a reasonable request. Asking for their insurance coverage is crucial. If the subcontractor is not properly insured and one of the employees is injured on your job, you may be liable for those injuries. You can customize your applications to suit your needs. It may be wise to discuss the form and content of your subcontractor applications with an attorney.

Basic Interviews

There are many questions you will want to ask in the initial interview with a subcontractor. When you conduct your interviews, you want to derive as much insight into the qualities of the subcontractors as possible. These interviews will be the basis for your decision to hire or not to hire subcontractors.

If you have a professional office, that is the best place to meet subcontractors. If your office conditions don't reflect the image you want to give, meet the subcontractors on neutral ground. You could meet them in a coffee shop, restaurant, or other place.

PRO POINTER

There are many questions you will want to ask in the initial interview with a subcontractor. When you conduct your interviews, you want to derive as much insight into the qualities of the subcontractors as possible. These interviews will be the basis for your decision to hire or not to hire subcontractors.

Pick a meeting place that will allow you to project your best image. One of the best places to meet your subcontractors is at their place of business. First of all, you'll impress them with the fact that you took the time to come to their office. You can ask for a tour of their shop to look at the way in which they store materials and equipment and possibly see one of their trucks being loaded to go on a job. In their office, you can talk to their bookkeeper, possibly their estimator if they have one, and get a general idea of their office and field operations. During the interview, set the pace and go through your list of questions. Let the subcontractor talk, but you should set the pace for the interview.

Checking References

Checking references should be standard procedure when selecting subcontractors. If a subcontractor has been in business for any length of time, he or she should have references. Ask for these references, and follow up on them. But remember that the subcontractor is likely to give you only good references. As you talk to each reference, ask for other companies that the subcontractor has had a business relationship with and then contact them.

PRO POINTER

Checking references should be standard procedure when selecting subcontractors. If a subcontractor has been in business for any length of time, he or she should have references. Ask for these references, and follow up on them. But remember that the subcontractor is likely to give you only good references. As you talk to each reference, ask for other companies that the subcontractor has had a business relationship with and then contact them.

Checking Credit

Another part of screening subcontractors is checking credit. By checking the credit ratings of subcontractors, you can determine a good deal about the individuals who own the company and some financial history of their business.

If subcontractors have had bad credit, it may not mean that they should be rejected as a potential hire, since there may be extenuating circumstances. I've known a few subcontractors who were not paid by general contractors who skipped town, owing lots of people. As a result these subs had difficulty paying their bills on time for quite a while. They finally absorbed these big losses and got back on track, but someone just looking at the bad credit report without knowing all the facts would form the wrong opinion of these two reliable companies. Credit reports can tell you a lot about the people you will be doing business with but not everything.

Read between the Lines

In all your business endeavors you must learn to read between the lines. Credit reports are a good example of where the facts may not tell the whole story. Let's say you are reviewing a credit report and see that a subcontractor has filed for bankruptcy; would you subcontract work to this individual? Look further—you may be missing out on a good subcontractor. The fact that someone has filed for bankruptcy is not enough to rule out doing business with the individual. Individuals can get into financial trouble through no fault of their own, as the above example shows. You must be willing and able to decipher what you are seeing. When you learn to read between the lines, you will be a more effective businessperson.

Set the Guidelines

If you plan to utilize the services of the subcontractors you are interviewing, set the guidelines for doing business with your firm. If you require all your subcontractors to carry pagers, tell them so. If you require subcontractors to return your phone calls within an hour, make the point clearly. Remember that you are in control, but you can't expect people to read your mind. You have to let your desires be known.

PRO POINTER

If you plan to utilize the services of the subcontractors you are interviewing, set the guidelines for doing business with your firm. If you require all your subcontractors to carry pagers, tell them so. If you require subcontractors to return your phone calls within an hour, make the point clearly. Remember that you are in control, but you can't expect people to read your mind. You have to let your desires be known.

Coming to Terms

Coming to terms is a key issue in selecting subcontractors. What you want and what the subcontractors want may not be the same. If you are going to do business with subcontractors, you should work out the terms of your working arrangements in advance.

Discussing contracts is a vital topic in meeting with subcontractors. You should go over your subcontract agreements with the subcontractors. If there are any questions or hesitations, resolve them in the meeting. You don't want to get into the middle of a job and find out that your subs will not play by the rules.

The more detail you go into in the early stages of your relationships, the more likely you are to develop good working conditions with your subcontractors. Just like your contracts with homeowners, you want your subcontractor agreements to be free of confusion and misinterpretation. Insert whatever clauses are appropriate to make sure subs understand what you expect. Take as much time as necessary, but remove any doubts from your contracts.

The contract should include the time (date) when you expect the sub to be on a job; also include the time (date) when their work is to be completed (and accepted as satisfactory). If the sub fails to meet these dates, what leverage do you have? As far as starting date, you don't have much, but if the contractor delays too much, you'll need to send notification that if the contractor fails to start on such-and-such date, you will cancel the contract. This is the reason for having multiple subs in each trade ready to do business with you.

As far as completion is concerned, you have a little more leverage. A standard contract provision, known as a "performance clause," deals with this issue. A typical performance clause is similar to the following:

> *The subcontractor agrees to commence and complete the subcontractor's work by (dates) and to perform the work at lesser or greater speeds, and in such quantities as, in the contractor's judgment is required for the best progress of the job or as specifically requested by the contractor. Should the subcontractor fail to prosecute the work or any part thereof with promptness and diligence or fail to supply a sufficiency of skilled workers or materials of proper quality, the contractor shall be at liberty after seventy-two hours' written notice to the subcontractor to provide such labor and materials as may be necessary to complete the work and to deduct the cost and expense thereof from any money due or thereafter to become due to the Subcontractor under this agreement.*

Another subject for the contract is job cleaning. Subcontractors are generally required to clean up their trash and debris, but many don't do it promptly, and many don't do it at all unless there is a clause in the contract such as:

> *The subcontractor shall at all times keep the job site clean and orderly and free from dirt arising out of the subcontractor's work. At any time, upon the Contractor's request, subcontractor shall immediately clean up and remove from the job site anything it is obligated to remove hereunder or Contractor may, at its discretion and without notice, perform or cause to be performed such cleanup and removal at the Subcontractor's expense.*

Payment issues are always important. In the industry, there is widespread use of a "pay when paid" clause in many subcontract agreements. Some states have deemed these clauses invalid, and you may not wish to use them. If you do, the clause typically states that the subcontractor will be paid within "X" days (generally 15 to 30 days) after the contractor has received payment for that work from the owner. In effect, this pegs your payment to the subcontractor to payment from the owner. If you are paid promptly, your subcontractor will be as well; however, if the owner delays payment to you, you can delay that corresponding payment to your subcontractor.

Delays frequently occur in this business, and they often affect some of your subcontractors, who share no responsibility in those delays. However a sub may look to the general contractor for reimbursement of costs they incurred because of these delays. These costs are referred to as "consequential damages." A clause in the subcontract agreement stipulating that "the contractor shall not be liable to the subcontractor for any damages or extra compensation that may occur from delays in performing work or furnishing materials or other causes attributable to other subcontractors, the owner, or any other persons" will do the trick.

Maintaining the Relationships

Once you hire new subcontractors, you must concentrate on maintaining the relationships. When you find good subs, you have probably invested a significant amount of time in their selection, and to avoid wasting this time you must work on building these relationships. This doesn't mean that you have to become buddies with your subs, but you do have to fulfill your commitments and expect them to fulfill theirs.

If you tell a subcontractor that you will pay bills within five days of receiving them, you had better be prepared to pay the bills. When you agree to terms in your subcontractor agreements, stick to them. If you breach your agreements with subcontractors, they will at some point decide to go elsewhere with their business.

Rating Subcontractors

Rating subcontractors will take some time and effort, but it will be worth it. You should rate subs in order to develop the best team possible. This rating procedure starts with the interview but doesn't stop there. You should look at many aspects of the subcontractor's business and history.

One of the first qualities you should evaluate is the work history of the subcontractor. Experience counts, not only in terms of the company but in the people you'll be working with. For example, a person with 15 years of experience who has just gone into business may be a better choice than one who has been in business for two years but only has five years of experience. There are advantages to choosing a subcontractor who has an established business.

If the subcontractor has been in business for awhile, there is a better chance that the business will last. New businesses often fail within the first two years, but businesses that have been around for between three to five years have a better chance of survival. Business owners who survive these early years have business experience and dedication; these are the traits you should look for in a subcontractor.

The business practices employed by subcontractors can affect their desirability. This aspect may be difficult to assess in a single meeting, but by doing some research you can learn much about how the sub conducts business. You need a subcontractor who is responsive to your phone calls. With beepers and cell phones quick communication is available to nearly everyone. Most subs will assure you that they are attentive to answering calls, but you need to verify their claims. A quick test can be conducted shortly after the subcontractor leaves your office; call him. You know that he has just left the meeting and is not in his office, and you will find out how his phone is answered and how quickly he will return your call.

Before you commit to using a subcontractor, conduct a test. Call three contractors and schedule a bid meeting. Be prepared to discuss a pending job; if you don't have any, create a simple project that you are going to ask them to bid. You will be looking for certain key responses. Will they be punctual in responding? How long will it take to get the quotes you want? Will their response be professional and clear in the scope of work included or excluded? Will they offer any suggestions to improve quality or price?

If you created a "project," after you have received all their responses and the test is over, you can merely tell them that the job didn't go ahead and thank them for responding. You have now found out how each subcontractor would have responded to a real situation.

Tools and equipment are another consideration when judging subcontractors. If your subcontractors don't have the necessary tools and equipment, they will not be able to give you the service you desire. Don't hesitate to inquire about the tools and equipment owned by the subcontractor.

Insurance coverage is a very important subject. You cannot afford to use subcontractors who are not prop-

> **PRO POINTER**
>
> Insurance coverage is a very important subject. You cannot afford to use subcontractors who are not properly insured. It is easy to lose your business in a lawsuit concerning personal or property damage. For your own protection, you must make sure your subcontractors carry the proper amounts of personal liability and property insurance.

erly insured. It is easy to lose your business in a lawsuit concerning personal or property damage. For your own protection, you must make sure your subcontractors carry the proper amounts of personal liability and property insurance.

If an employee of the subcontractor is injured on your job and the sub doesn't have insurance, you could be liable for any claims that that employee may file. For this and other reasons you need to have your subcontractors submit certificates of insurance to you before starting work on your project.

If the sub has employees other than close family members, worker's compensation insurance will be needed. Even if the subcontractor is not required to carry worker's comp, you should have a waiver signed by the business owner. The waiver, sometimes referred to as a "hold harmless" clause, should be prepared by your lawyer, so you can be protected from any potential claim.

If you use the services of subcontractors who are not properly insured, you may have to pay up at the end of the year. When your insurance company audits you, as they usually do, you will be responsible for paying penalties if you used improperly insured contractors. These penalties can amount to a substantial sum of money. To avoid losing money, make sure your subcontractors are currently insured for all necessary purposes.

Subcontractor Specialization

Many subcontractors specialize in various areas of work. Some HVAC contractors only fabricate ductwork; others install equipment; others may only install low-voltage control wiring. Some electrical contractors specialize in low-voltage data and voice communication wiring and installations. When you are dealing with specialists, you may pay extra, but the end result may be a bargain. While specialists may charge higher fees, they are often worth the extra cost because they can complete their work more quickly, allowing you to bring in the next trade sooner. Remember that time is money, and when you save time you have a chance to make more money. With this in mind, ask potential subcontractors what they specialize in. You may find it cost-effective to use different subs for different jobs.

Licenses

Licenses are another issue you should investigate when rating subcontractors. If subcontractors are not licensed as required by the local or state authorities, you can get into deep trouble. It is important that you only hire subcontractors that meet and have obtained standard licensing requirements. If you use unlicensed subcontractors, you are flirting with disaster.

Work Force

A subcontractor's work force is another consideration. How much work can the contractor handle? You don't want to hire a sub who cannot handle your workload. For this reason, you must know what the capabilities of the subcontractors are.

I've rarely heard a subcontractor refuse new work, even when he already has a full workload. Subcontractors who take on more work than they can handle will surely disappoint some builder—and it could be you. They will promise to be on the site, but, because of another contractor's demands that day, they'll go to the other site. You need to investigate how much work the subcontractor has and whether or not he can honestly service your job when needed.

If you give a small contractor too much work, you will find yourself in a bind. The small contractor may hire either temporary employees or inexperienced or substandard workers to keep your business.

While it is more convenient to work with a single subcontractor, it may not be practical due to his current workload, so you need to develop a core of subs in the same trade. As a safety precaution, you should have at least three subcontractors in each trade. This depth of subcontractors will give you more control.

It is also good to introduce competition among subcontractors. By using the same subcontractor over and over, other subs may be reluctant to bid on future work, assuming you always favor the one you have continued to hire, so you may not get truly competitive pricing.

> **PRO POINTER**
>
> Licenses are another issue you should investigate when rating subcontractors. If subcontractors are not licensed as required by the local or state authorities, you can get into deep trouble. It is important that you only hire subcontractors that meet and have obtained standard licensing requirements. If you use unlicensed subcontractors, you are flirting with disaster.

> **PRO POINTER**
>
> While it is more convenient to work with a single subcontractor, it may not be practical due to his current workload, so you need to develop a core of subs in the same trade. As a safety precaution, you should have at least three subcontractors in each trade. This depth of subcontractors will give you more control.

Managing Subcontractors

Managing subcontractors will be much easier when you follow some simple rules. The most important rule is to document your dealings in writing. Other rules include:

- Create and use a subcontractor policy.
- Be professional and expect professionalism from the subs.
- Use written contracts with all of your subcontractors.
- Use change orders for all deviations in your agreement.
- Dictate start and finish dates in your agreement.
- Back-charge subcontractors for costs due to poor performance or poor quality.
- Always have subs sign lien waivers when they are paid.
- Keep certificates of insurance on file for each sub.
- Don't allow extras unless they are agreed to in writing.
- Don't give advance contract deposits.
- Don't pay for work that hasn't been inspected.
- Use written instruments for all your business dealings.

Subcontractors can take advantage of you if you let them. However, if you establish and implement a strong subcontractor policy, you should be able to handle your subs. It is imperative that you remain in control. If subcontractors have the lead role, your company will be run by the subs not by you.

Dealing with Suppliers

Dealing with suppliers is not as simple as placing an order and waiting. Your business depends on the performance of suppliers, and it is up to you to set the pace for all your business dealings. Establish fair rules and make sure they follow them. Be firm but fair.

Establish a routine with your suppliers. If you are going to use purchase orders, use them with every order.

PRO POINTER

Dealing with suppliers is not as simple as placing an order and waiting. Your business depends on the performance of suppliers, and it is up to you to set the pace for all your business dealings. Establish fair rules and make sure they follow them. Be firm but fair.

When you want the job name and address written on your receipts, insist that they are always included. Are you going to allow employees to make purchases on your credit account? If so, set limits on how much can be purchased, and make sure everyone at the supply house knows which employees are authorized to charge on your account.

Get to know the manager of the supply house. Without a doubt, there will come a time when you and the manager will have a problem to solve. At these times it helps to know each other.

When you use a new supplier, make sure you understand the house rules. What is the return policy? Will you be charged a restocking fee? Will you get a discount if you pay your bill early? What is your discount percentage? Will the discount remain the same regardless of the volume you purchase? These are just some of the questions you should get answers to.

If all goes well, you will be doing a lot of business with your suppliers. Since each of you depends on the other for making money, you should develop the best relationship possible.

Making Your Best Deal

How will you know when you are making your best deal? Is price the only consideration in the purchase of materials or the selection of subcontractors? No; service and quality are two other important factors in that equation. Getting the lowest price doesn't always mean you are getting the best deal. If you don't get quality and service to go along with a fair price, you are probably asking for trouble. Let me give you a few examples.

Assume that you have requested bids from five painters. You accept the lowest bid based on price alone. When the painters are scheduled to start the job, they don't show up. After calling and insisting that they be on the job by the next morning, you get some satisfaction. The painters show up and start to work. You go back to your office and at

PRO POINTER

How will you know when you are making your best deal? Is price the only consideration in the purchase of materials or the selection of subcontractors? No; service and quality are two other important factors in that equation. Getting the lowest price doesn't always mean you are getting the best deal. If you don't get quality and service to go along with a fair price, you are probably asking for trouble.

noon the homeowner calls, wanting to know where the painters are. You find out the painters left for a morning break and never came back.

This does not make your company look good in the eyes of your customer. What does the slowdown do to your cash flow? It crimps it, of course. You got the cheapest painter you could find, and your great deal doesn't look so good now. This type of problem is common, and you will have to do a better job of finding suitable subcontractors in the future.

For the next example, assume that you have ordered roof trusses from your supplier. After shopping prices, you decided to go with the lowest price, even though you had never dealt with the supplier before. The trusses are ordered and you are given a delivery date. All your work is scheduled around the delivery of the trusses.

If the trusses are to be used to replace a rotted roof structure, you can't tear off the old roof until you know the trusses are available. On the day of delivery, you call the supplier and inquire about the status of the trusses. You're told the trusses are on a delivery truck and will be on your job by mid-morning.

Your crew finishes framing work and is waiting to set trusses. It's nearly noon, and the trusses have not yet been delivered. Your crew is at a standstill. A phone call to the supplier reveals that the delivery truck broke down on the way to the job. (The broken-truck ploy is an old one; perhaps the truck did break down, but more likely the supplier messed up your delivery and is looking for an excuse.) You're told the trusses won't arrive until the next morning, but now what are you going to do? You've already lost money while your crew was idle waiting for the trusses.

When you ask the supplier to transfer the trusses to a different delivery truck so you can get them immediately, you're told that the supplier doesn't have another truck capable of transporting the trusses. As it turns out, you have to leave the job until the trusses arrive, losing both time and money. And you have really upset your customer.

Would this have happened if you had used your regular supplier? Probably not, because that supplier has enough trucks to make a switch if necessary. Your great deal on inexpensive trusses has turned into a disaster. So, you see, price isn't everything.

Expediting Materials

Learning how to expedite materials will keep your business running on the fast track. Work cannot get done unless there are materials to work with. If there is no work, there is no money coming in but lots of money going out. Since business owners are in business to make money, they need to keep materials available and flowing.

Placing an order and periodic follow-up calls takes a fair amount of time and effort, but if a worker has to leave the job to go pick up materials at a supply house, time and money are lost. Inaccurate takeoffs and inattention to deliveries can cost contractors thousands of dollars. Is there anything that can be done to reduce these losses? Yes, by expediting materials, more time is saved and more money is made.

All too many contractors call in a material order and forget about it. They don't make follow-up calls to check the status of the material. It is not until the material doesn't show up that these contractors take action. By then, time and money are being lost.

Many contractors never inventory materials when they are delivered. If 100 sheets of plywood were ordered, they assume that they received 100 sheets of plywood. Unfortunately, mistakes are frequently made with material deliveries. Quantities are not what they are supposed to be. Errors are made in the types of materials shipped. All these problems add up to more lost time and money.

When you place an order, have the order taker read the order back to you. Listen closely for mistakes. Call in advance to confirm delivery dates. If a supplier has forgotten to put you on the schedule, your phone call will correct the error before it becomes a problem.

When materials arrive, check the delivery for accuracy. Ideally, this should be done while the delivery driver is present. If you discover a fault with your order, call the supplier immediately. By catching blunders early, you

PRO POINTER

Learning how to expedite materials will keep your business running on the fast track. Work cannot get done unless there are materials to work with. If there is no work, there is no money coming in but lots of money going out. Since business owners are in business to make money, they need to keep materials available and flowing.

PRO POINTER

Keeping a log of material orders and delivery dates is one way of staying on top of your materials. One glance at the log will let you know the status of your orders. When you talk to various salespeople, record their names in your log. If there is a problem, it always helps to know whom you talked to last. Get a handle on your materials, and you will enjoy a more prosperous business.

can reduce your losses. And don't forget to note any deficiencies or discrepancies on the delivery ticket so there is a written record of the problem.

Keeping a log of material orders and delivery dates is one way of staying on top of your materials. One glance at the log will let you know the status of your orders. When you talk to various salespeople, record their names in your log. If there is a problem, it always helps to know whom you talked to last. Get a handle on your materials, and you will enjoy a more prosperous business.

Avoiding Common Problems

By avoiding common supplier and subcontractor problems, you can spend more time making money. The two biggest reasons for problems between contractors and suppliers or subcontractors are poor communication and money. Money is usually the major cause of disputes, and communication breakdowns cause the most confusion. If you can conquer these barriers, your business will be more enjoyable and more profitable.

There are few excuses for problems in communication if you always use written agreements. When you give a subcontractor a spec sheet that calls for a specific make, model, and color, you eliminate confusion. If the subcontractor doesn't follow the written guidelines, an argument may ensue, but you will be the victor.

As for money, written documents can solve most of the problems caused by cash. When you have a written agreement that details a payment schedule, there is little for anyone to get upset with. By using written agreements, you can eliminate most of the causes for disagreements and arguments. It's a good idea to create a boilerplate bid form to use in conjunction with plans and specifications. When these forms are designed for specific trades, you can eliminate confusion and mistakes during the bidding process.

Learning to work well with subcontractors and suppliers is essential to the success of a building business. Work hard to develop and keep good relationships. You depend on a team of people when you build homes for a living, so you need to concentrate on and develop team-building qualities

Job Name: _____

Phase: Foundation

Contractor: _____

All Work To Be Done According To Attached Specifications

Bid Item

Supply labor and material for footings

Supply labor and material for foundation walls and piers

Supply and install foundation windows/vents

Supply labor and material to create bulkhead opening, ready for door installation

Supply and install foundation bolts

Remove all foundation clips

Waterproof foundation to finished grade level

Supply labor and material to install concrete basement floor

FIGURE 10-1   Foundation bid sheet.

Job Name: _____

Phase: Paint

Contractor: _____

All Work To Be Done According To Attached Specifications

Bid Item

Provide price for labor and material to paint, stain, and/or seal all surfaces specified

Price should include all preparation work required (i.e., filling nail holes)

FIGURE 10-2   Paint bid sheet.

Job Name: _____

Phase: Drywall

Contractor: _____

All Work To Be Done According To Attached Specifications

Bid Item

Supply and install all materials needed to drywall all interior walls and ceiling to code requirements

Provide separate labor only price for hanging, taping, and finishing drywall

If heat is needed, drywall contractor shall supply it

Provide separate price for texturing ceilings

FIGURE 10-3   Drywall bid sheet.

Job Name: _____

Phase: Heating

Contractor: _____

All Work To Be Done According To Attached Specifications

Bid Item

Supply and install all rough heating materials and finished heating equipment including boiler and baseboard units

FIGURE 10-4 Heating bid sheet.

Job Name:
_____

Phase: Tree Clearing

Contractor:
_____

All Work To Be Done According To Attached Specifications

Bid Item

Cut all trees marked with blue ribbons
Remove all wood, branches, brush, and debris from cutting procedure.

FIGURE 10-5 Tree clearing bid sheet.

Job Name: _____

Phase: Plumbing

Contractor: _____

All Work To Be Done According To Attached Specifications

Bid Item

    Supply and install all rough plumbing and plumbing fixtures, including bathing units

    Bid a separate price for well pump and related equipment

FIGURE 10-6   Plumbing bid sheet.

Job Name: _____

Phase: Well

Contractor: _____

All Work To Be Done According To Attached Specifications

Bid Item

    Supply labor and material to install drilled well with steel casing and cap

    Supply labor and material to install submersible pump and related equipment

    Bid job on a per-foot basis and on a flat-fee basis

FIGURE 10-7   Well bid sheet.

Job Name: _____

Phase: Siding

Contractor: _____

All Work To Be Done According To Attached Specifications

Bid Item

  Install siding materials provided by general contractor

FIGURE 10-8   Siding bid sheet.

Job Name: _____

Phase: Framing

Contractor: _____

All Work To Be Done According To Attached Specifications

Bid Item

  Supply labor to frame house to a dried-in condition
  If a crane is needed, it will be at the framing contractor's expense
  Install all exterior windows and doors
  Subfloors are to be glued and nailed
  Provide access for bathtubs and showers
  Install ceiling strapping
  Install all steel beams and plates as might be required
  Install all support columns
  Build stairs during initial framing

FIGURE 10-9   Framing bid sheet.

Job Name: _____

Phase: Flooring

Contractor: _____

All Work To Be Done According To Attached Specifications

Bid Item

Provide price for supplying and installing underlayment
Provide price for labor and material to prepare all floor surfaces
Provide price to supply and install flooring as specified

FIGURE 10-10   Flooring bid sheet.

Job Name: _____

Phase: Insulation

Contractor: _____

All Work To Be Done According To Attached Specifications

Bid Item

Supply and install all insulation

FIGURE 10-11   Insulation bid sheet.

Job Name: _____

Phase: Trim

Contractor: _____

All Work To Be Done According To Attached Specifications

Bid Item

Supply labor to install trim materials supplied by general contractor
Provide separate price for installing counters and cabinets in all areas
Trim price should include hanging all interior doors, installing window and
door hardware, and bath accessories

FIGURE 10-12   Trim bid sheet.

Job Name: _____

Phase: Roofing

Contractor: _____

All Work To Be Done According To Attached Specifications

Bid Item

Install roofing materials provided by general contractor

FIGURE 10-13   Roofing bid sheet.

Job Name: _____

Phase: Electrical Work

Contractor: _____

All Work To Be Done According To Attached Specifications

Bid Item

Supply and install temporary power pole
Supply and install all rough wiring
Install light fixtures supplied by general contractor
Electrical contractor to provide GFI devices and smoke detectors
Supply all needed permits and inspections

FIGURE 10-14   Electrical work bid sheet.

Job Name: _____

Phase: Site Work

Contractor: _____

All Work To Be Done According To Attached Specifications

Bid Item

    Remove all tree stumps and debris from any excavation
    Supply and install metal culvert pipe for driveway
    Install driveway—site contractor to furnish all materials
    Dig foundation hole
    Provide rough grading
    Backfill foundation
    Perform final grading
    Seed and straw lawn
    Install septic system
    Supply and install foundation drainage
    Supply and install crushed stone for foundation
    Dig trenches for water service and sewer
    Backfill trenches for water service and sewer

FIGURE 10-15   Site work bid sheet.

Preparing Winning Bids

To get jobs, you often need to win a bidding war with other builders. Occasionally, you can ride your reputation to a job without competition. Normally, you will be up against competitive contractors in your quest for work. Being successful as a builder is directly related to winning bids.

What is your strategy? Are you going to try to be the low bidder? This is usually not a good idea. How about being the high bidder? Probably won't work either. Most winning bids are somewhere in the middle of the financial offerings. So, what makes one bid win over another when they are close in the numbers? Many factors can come into play. Here are a few of them:

- Presentation
- Reputation
- Experience
- Exposure
- Name recognition
- References
- Track record

Knowing how to get work is essential, and that is our topic here.

Word-of-Mouth Referrals

Word-of-mouth referrals are the best way to get new business. Of course, you will need some business before you can benefit from word-of-mouth referrals. But every time you get a job, you need to think about using that customer as a referral for future contracts. A satisfied customer is your best route to attracting new clients.

Getting referrals from existing customers is not only the most effective way to generate new business; it's also the least expensive. Advertising is expensive. For every job you get from advertising, you are losing a percentage of your profit to the cost of advertising. If you can develop new work from existing customers, you eliminate the cost of advertising.

> **PRO POINTER**
>
> Word-of-mouth referrals are the best way to get new business. Of course, you will need some business before you can benefit from word-of-mouth referrals. But every time you get a job, you need to think about using that customer as a referral for future contracts. A satisfied customer is your best route to attracting new clients.

If you do quality work and live up to your commitments to your customers, referrals will be easy to come by—but you may have to ask for them. People will sometimes give your name and number to friends, and they occasionally write nice letters. However, to make the most of word-of-mouth referrals, you have to learn to ask for what you want. Let's see what it takes to get the most mileage out of your existing customers.

Laying the groundwork is an important step in getting a strong portfolio of customer referrals. If you don't make your customers happy, they will talk to their friends, but they won't be saying too many kind words about you or your company. People are quick to talk about their bad experiences, but they may not be so quick to spread good words. To get the message out that you want referrals, you have to work hard at pleasing your customers.

Start with the first contact you have with customers and work to continue that relationship. Call them from time to time after you have completed the work to find out how they are enjoying that new addition or the new bathroom or new house. Ask about the family, how the kids are doing. This show of interest will not be lost on your customers, and when asked for a referral they will respond with a positive one.

I have seen many good jobs go bad in the final days of their completion. One of the biggest mistakes a contractor can make is not responding promptly to warranty calls. All the goodwill you have created throughout the job may go down the drain because you failed to respond promptly to that leaking faucet, that toilet that won't stop running, or that bedroom door that won't close properly. If you are in business for the duration, you want to please your customers, not alienate them, particularly after the job is done. Old customers become repeat customers. If you don't respond quickly and effectively to callbacks, you won't be called when there is new work to be done.

During the job you must cater to the customers by being punctual at all your meetings, having answers to their questions, listening to any problems they may have observed, being respectful, and being professional. Always keep in the back of your mind the thought that you want these customers to be happy with your work because you'd like to use them as a reference.

At the completion of the job you should ask your customers for a reference; don't expect them to offer to give you one. Often, asking for a letter of reference isn't enough. It helps if you provide a form for the consumer to fill out. People never seem to know what to say in a reference letter. They are much more comfortable filling out forms.

> **PRO POINTER**
>
> At the completion of the job you should ask your customers for a reference; don't expect them to offer to give you one. Often, asking for a letter of reference isn't enough. It helps if you provide a form for the consumer to fill out. People never seem to know what to say in a reference letter. They are much more comfortable filling out forms.

If you design a simple form, almost all satisfied customers will complete and sign it. I'm sure you have seen these quality-control forms in restaurants and with mail-order shipments. You can structure the form in any fashion you like.

Once you have created and printed your forms, use them. When you are completing a job, ask your customer to fill out and sign your reference form. Do it on the spot. Once you are out of the house, getting the form completed and signed will be more difficult.

As you begin building a collection of positive reference forms, don't hesitate to show them to prospective customers. Use an attractive three-ring binder and clear protective pages to store and display your hard-earned references. When you get enough reference letters, you will have strong ammunition to close future deals.

Customer Satisfaction

Customer satisfaction is the key to success. Business builds upon itself when customers are satisfied. Of course, there are some people whom you may never be able to satisfy, and these hard-to-please people seem to exist for every business owner, but they are the exception rather than the rule. Concentrate on pleasing the bulk of your customers.

Customers like to feel comfortable with their contractors. To make customers comfortable, you have to deal with them on a level that they feel at ease with. Remember, they may not be familiar with construction or construction terms, so you have to speak to them in terms that they understand. Communication skills are essential in building good relationships. If you and the customers can't communicate easily, you'll have a much more difficult time in your future business dealings. Some customers will require a lot of hand holding, and on occasion you may have to spend more time with these clients in order to make them feel comfortable with the decisions they have made or are planning to make.

PRO POINTER

Customer satisfaction is the key to success. Business builds upon itself when customers are satisfied. Of course, there are some people whom you may never be able to satisfy, and these hard-to-please people seem to exist for every business owner, but they are the exception rather than the rule. Concentrate on pleasing the bulk of your customers.

Reaching Out for a New Customer Base

You can start reaching out for a new customer base through the use of bid sheets, a system of notification of pending jobs in either the public or private sector that are advertised. Bid sheets are a way of notifying contractors of potential projects, and any contractor can bid on them. Listing a project on a bid sheet is generally recognized as a commitment by the owner to proceed with the work if bids come in within their budget, unlike some estimates requested by residential owners that may be requested for information purposes only and do not represent a real job. This type of competitive bid work frequently results in a job with a low profit margin, but if the estimate is correct it can still pay the bills.

Where Are Bid Sheets Available?

Bid sheets can be obtained by responding to public notices in newspapers and by subscribing to services that provide bid information. If you watch the classified section of major newspapers, you will see advertisements for jobs going out for bids. You can receive bid packages by responding to these advertisements. Normally, you will get a set of plans, specifications, bid documents, bid instructions, and other needed information. These bid packages can be simple or complicated, depending upon the nature and size of the project.

PRO POINTER

Bid sheets can be obtained by responding to public notices in newspapers and by subscribing to services that provide bid information. If you watch the classified section of major newspapers, you will see advertisements for jobs going out for bids. You can receive bid packages by responding to these advertisements.

Subscription services are available both online and via postal mail. These services will provide regular listings of jobs going out to bids. I have been in the business long enough to have used all types of services. I find the online services to be the most efficient.

What Is a Bid Sheet?

A bid sheet is a formal request for price quotes. A typical bid sheet contains bidding information similar to this:

Requests for Bids:

A 3,500-square-foot, two-story new residence is being planned for a 1/2-acre parcel located on the corner of St. Paul and Charles Streets, Baltimore, Maryland. Interested contractors can obtain the plans and specifications by contacting Mr. James Smith, architect, at the office located at 555 New Bern Avenue, Towson, Maryland, telephone: 410-854-4444. A refundable check in the amount of $175 is required for the bid package. Bids are due on April 15, 2009.

There is a difference between a bid sheet and a bid package. The bid sheet will give a brief description of the work available. A bid package gives complete details of what will be expected from bidders. Most contractors check bid sheets, and if they find a job of interest, they order a bid package. Bid sheets can be either free if posted in a newspaper or magazine or require a check if provided by a private service.

PRO POINTER

There is a difference between a bid sheet and a bid package. The bid sheet will give a brief description of the work available. A bid package gives complete details of what will be expected from bidders. Most contractors check bid sheets, and if they find a job of interest, they order a bid package. Bid sheets can be either free if posted in a newspaper or magazine or require a check if provided by a private service.

Bid-Reporting Companies

There are several companies that are in the business of providing bid information. Local addresses of these types of companies can be found in most Yellow Pages telephone directories or over the Internet. These listings are normally published in a newsletter form. The bid reports are generally delivered to contractors on a weekly basis. Each bid report may contain five jobs or fifty jobs. These publications are an excellent way to get leads on all types of jobs.

All types of projects are included in these bid sheets. They range from small residential jobs to large commercial jobs. The majority of the jobs are commercial. The size of the jobs ranges from a few thousand dollars to millions of dollars.

Public-Agency Bid Sheets

Bid sheets prepared by local, state, or federal government agencies present another opportunity to find work. Like other bid sheets, government sheets give a synopsis of the job description and provide information for obtaining more details. Government jobs can range from replacing a dozen lavatory faucets to building a commissary. Building new base housing units for the armed forces could result in many months of work for a homebuilder.

Government agencies represent a fairly stable type of project; however, there is a lot of paperwork, and sometimes payments are slow in coming. There are several things to consider when deciding to bid on a government project: There will be lots of forms to fill out and lots of paperwork, a bond will be required, and there are many laws and ordinances that will have to be followed. If you are not willing to deal with mountains of paperwork, stay away from government bids.

If public funds are used in any of these types of projects, they will fall under the state and federal affirmative action and equal-opportunity laws. If you qualify, provisions in these types of contracts to eliminate discrimination may be beneficial to you. A typical clause in those bid documents would be similar to this one:

> *For the purpose of this contract, a goal of "X" percent has been established for socially and economically disadvantaged businesses that are owned and controlled by those individuals who are Black Americans, Hispanic Americans, Asian-Pacific Americans, Subcontinent Americans, Native Americans, or Women, pursuant to [the laws].*

A company certified by the state and/or federal government as an MBE (Minority Business Enterprise) or a WBE (Women-Owned Business Enterprise) is encouraged to bid, and a specific number of these firms will be hired for portions of the project. If you feel you can qualify for any of these designations, a little investigation might be beneficial to you if you wish to seek work in the public sector.

Payment, Performance, and Bid Bonds

Payment, performance, and bid bonds are a necessity with many major jobs. If you order a bid sheet or package, you will almost certainly see that a bond is required. Some listings on bid sheets may not require a bond. Requirements for bonds are often tied to the anticipated cost of the job; the bigger the job, the more likely it is that a bond will be required. All public projects will require bonds.

Are You Bondable?

Many of the jobs listed on these bid sheets require contractors to be bonded. Bonds are obtained from bonding and insurance companies. Before you try bidding jobs that require bonding, check to see if you are bondable. The requirements for being

PRO POINTER

Many of the jobs listed on these bid sheets require contractors to be bonded. Bonds are obtained from bonding and insurance companies. Before you try bidding jobs that require bonding, check to see if you are bondable.

bonded vary. Check in your local phone book for an insurance agent who also handles bonds for contractors, and call to inquire about the requirements.

Why Are Bonds Required?

Bonds are required to ensure the successful completion of a project. A payment bond assures the owner of the project that all suppliers, vendors, and subcontractors will be paid at the end of the project. A performance bond assures the owner that the contractor will "perform" in accordance with the terms and conditions of the contract. If the contractor defaults on either of these obligations, the bonding company steps in and provides the necessary funds to satisfy these defaults.

When the person or firm awarding the contract requires a bond, they know there is a degree of safety. Bid bonds are slightly different. Bid bonds are generally required in an amount equal to 10 percent of the project's cost. They are submitted with the contractor's bid. If the contractor is notified that he is the successful bidder but declines to accept a contract, the owner will make the award to the second bidder. The proceeds from the bond will be used to make up the difference if any between the low bidder's price and the second bidder's price. The contractor who declined to accept the bid will lose his bond and forfeit the money it cost.

Thus, if you are required to submit a bid bond, make sure that you are willing to accept a contract if offered or else you'll pay the penalty.

It is very difficult for some new businesses to obtain a bond. If the new company doesn't have strong assets or a good track record, getting a bond is tough. But when you put up a bond, the value of the bond is at risk. If you default on your contract, you lose your bond to the person who contracted you for the job. Since many people use the equity in their home as collateral for a bond, they could lose their house. Bonds are serious business. If you can get a bond, you have an advantage in the business world.

PRO POINTER

Bonds are required to ensure the successful completion of a project. A payment bond assures the owner of the project that all suppliers, vendors, and subcontractors will be paid at the end of the project. A performance bond assures the owner that the contractor will "perform" in accordance with the terms and conditions of the contract. If the contractor defaults on either of these obligations, the bonding company steps in and provides the necessary funds to satisfy these defaults.

Big Jobs—Big Risks?

Do big jobs carry big risks? You bet they do. There are risks in all jobs, but big jobs do carry big risks. Should you shy away from big jobs? Maybe, but if you go into the deal with the right knowledge and paperwork, you should survive and possibly prosper.

Cash-Flow Problems

Cash-flow problems are looming when contractors take on big jobs. Unlike small residential jobs, big jobs don't generally allow for contractors to receive cash deposits. If you tackle these jobs, you will have to work with your own money and use your credit lines. When involved in a big project, each request for payment will be large and some will be huge. If payments are late and you have to pay out thousands of dollars to suppliers and subcontractors, they will only wait so long for their money before they pull off the job. You'll need the financial resources to pay these people if your customer is late with her payments. I once knew a very successful contractor who was given the opportunity to work on a multimillion-dollar project. He turned it down. When I asked him why, he said, "If these guys are late on one payment, it will drain my resources and hurt my other jobs—I think I'll pass." Wise decision. It's wonderful to think of signing a million-dollar job in your first year of business, but that job could put your business into bankruptcy court. Some lenders will allow you to use your contract as security for a loan, but don't bet your business on it. If you want to take on a big job, get your finances in order first.

PRO POINTER

Cash-flow problems are looming when contractors take on big jobs. Unlike small residential jobs, big jobs don't generally allow for contractors to receive cash deposits. If you tackle these jobs, you will have to work with your own money and use your credit lines.

Slow Pay

Slow pay is another problem, not necessarily related to big jobs, but you undoubtedly will run into slow payers during your business career. Keeping track of when your payments are due is extremely important when you have tackled a big job, but it can be equally important if you have a number of small jobs where the owners are lax in their payments. New businesses are especially vulnerable in succumbing to slow pay. The check you thought would come last month might not show up for another 6o or 9o days, so stay on top of your accounts receivable and don't be afraid to call that customer who is overdue in paying; he will probably think more of you as a businessperson who pays attention to business.

No Pay

It is not unknown for an owner to suddenly declare bankruptcy; while it is never good, it always seems to happen at the worst possible moment—just when you have submitted that big pay request. Knowing the financial situation of your customer is a first step in avoiding a "no pay" situation. You can request a copy of the owner's financial commitment from a bank or other lending institution. If privately financed, it becomes more difficult, but if you notice a trend toward slow payments, you'd better investigate. It may be nothing or it may be the beginning of a bankruptcy proceeding. Some contractors are reluctant to ask an owner for proof of financing, but many owners will appreciate your businesslike approach.

PRO POINTER

If you're working as a subcontractor, filing a mechanic's lien is the best course of action when your customer refuses to pay you. If as a general contractor you haven't been paid for labor or materials, a mechanic's lien can usually be levied against the property where the labor or materials were invested. If you have to file a lien, make sure you do it right. There are rules you must follow in filing and perfecting a lien. You can file your own liens, but I recommend working with an attorney on all legal matters.

If the subcontractors don't get paid, suppliers don't get paid. The ripple effect continues. Anyone involved with the project is going to lose. Some will lose more than others. Generally, when jobs go bad, the banks or lenders financing the whole job will foreclose on the property. These lenders normally hold a first mortgage on the property.

If you're working as a subcontractor, filing a mechanic's lien is the best course of action when your customer refuses to pay you. If as a general contractor you haven't been paid for labor or materials, a mechanic's lien can usually be levied against the property where the labor or materials were invested. If you have to file a lien, make sure you do it right. There are rules you must follow in filing and perfecting a lien. You can file your own liens, but I recommend working with an attorney on all legal matters. Here are some important items to remember:

- Check with your local government to find the exact requirements for filing a lien.

- You usually have to file a lien within a certain period of time after you last worked on the project, but be careful about this requirement. Let's say the time required to file the lien is 60 days after you have last worked on the project. Say it is day 59 and you haven't filed, so you go back to the project and clean out the strainer on the kitchen faucet. This won't work. Most laws state that you have to perform "meaningful work" within that time frame, and cleaning a strainer won't qualify.

- Since the lien will be filed against the owner of the property, you need to ensure that you have the correct name of the owner and the correct description of the property. This can generally be found in the land records at the Town Clerk's office.

The actual preparation and filing of a mechanic's lien is best left to your attorney. If the lien is not "satisfied" (paid by the owner), the court will allow a judgment to be filed against the property, and foreclosure will be the final step.

If you get any money, it will likely be a settlement in a reduced amount. You will never quite get the taste of that bad job out of your mouth, but you should learn from it. Did I let that first past-due payment go by without calling the owner? Were there signs that things were going bad, but I either didn't recognize them or thought that everything would turn out OK? Try to learn from your mistakes.

Completion Dates

When you sign a contract with a specific completion date, you had better be prepared to meet that date. Some contracts have penalty clauses for late delivery. Even if your contract does not penalize you, your reputation is on the line. When you commit to a completion date, make sure you make it.

> **PRO POINTER**
>
> When you sign a contract with a specific completion date, you had better be prepared to meet that date. Some contracts have penalty clauses for late delivery. Even if your contract does not penalize you, your reputation is on the line. When you commit to a completion date, make sure you make it.

Contracts for large projects frequently include liquidated damages, which are assessed for each day the project finishes beyond the date stated in the contract. The daily amount is meant to represent the cost to the owner for not having the project completed on time. These liquidated damages, referred to as LDs, can run from $50 per day to $5,000 per day or more. Liquidated damages and the possible loss of your bond can ruin your business. Contractors with limited experience on big and small jobs are often unprepared to project solid completion dates. Don't sign a contract with a completion date you are not sure you can meet.

The Bid Process

When you learn how to beat your competition in the bid process, you are on your way to a successful business. There is no shortage of competition in most fields of contracting. Some contractors can get discouraged. But don't—there are ways to forge ahead of the competition.

The Competitive Nature of the Bid Sheet

Beating the competition when projects are obtained from bid sheets is tough. Low bidder gets the job, and with three or four or sometimes a dozen bidders competing for the work, someone can make a mistake in his estimate or decide to submit a "lowball' bid to get the job for one reason or another. Being the low bidder can get you the job, but you may wish you had never seen the job. Don't bid a job too low. It doesn't do you any good to have work if you're not making money.

If you can get bonded, you have an edge. A lot of bidders can't. This fact alone can be enough to cull the competition. When you prepare your bid package for submission, be meticulous. If you want the job, spend sufficient time preparing a professional bid packet. Discuss the job with several qualified subcontrac-

PRO POINTER

Beating the competition when projects are obtained from bid sheets is tough. Low bidder gets the job, and with three or four or sometimes a dozen bidders competing for the work, someone can make a mistake in his estimate or decide to submit a "lowball' bid to get the job for one reason or another. Being the low bidder can get you the job, but you may wish you had never seen the job. Don't bid a job too low. It doesn't do you any good to have work if you're not making money.

tors to get the best price; shop for materials and equipment, and double-check your estimate to see if you left something out or included something you shouldn't have. Check for errors in your math. Then double-check it. Don't let a bid out of your office without being sure it is ready for public viewing.

Submitting Bids in Person

If you are dealing with bids delivered to the customer in person, follow the guidelines found throughout this book. Here are some of the basic keys to success:

- Dress appropriately.
- Drive an appropriate vehicle.
- Be professional.
- Be friendly.
- Get the customer's confidence.
- Produce photos of your work.
- Show off your letters of reference.
- Give your bid presentation in person.
- Follow up on all your bids.

Preparing Accurate Take-Offs

To prepare accurate bids, you must be adept at preparing accurate take-offs. It doesn't matter if you use a computerized estimating program or a pen and paper—you must be precise. If you miss items in the take-off and get the job, you will lose money. If you overestimate the take-off, your price will be too high. An accurate take-off is instrumental in winning a job.

What Is a Take-Off?

A take-off is a list of items required to do a job. Take-offs from blueprints or visiting the job site produce a list of everything you will need to do the job. Some estimators are wizards with take-offs, and others have a hard time trying to figure out all the materials and components needed. If you can't discipline yourself to learn how to do an accurate take-off, you will need to either hire someone who can or find an estimating service that will provide you with an accurate estimate for a fee.

Using Take-Off Forms

You can reduce your risk of errors by using a take-off form. If you use a computerized estimating program, the computer files probably already contain forms. You may want to customize the standard computer forms. Whether you are using standard computer forms or making your own forms, you must be sure that they are comprehensive.

Take-off forms should include every item required for the type of job being estimated. It's best if you create forms that list every expense that you might incur on a job. This improves your odds of reducing errors and omissions when figuring prices. In addition, there should be a place on the form that allows you to fill in blank spaces with specialty items.

Take-off forms alert you to items you might otherwise forget. However, don't just look for items on your form. It is very possible that a job might require something that you haven't yet put on the form. Forms help, but there is no substitute for thoroughness.

PRO POINTER

To prepare accurate bids, you must be adept at preparing accurate take-offs. It doesn't matter if you use a computerized estimating program or a pen and paper—you must be precise. If you miss items in the take-off and get the job, you will lose money. If you overestimate the take-off, your price will be too high. An accurate take-off is instrumental in winning a job.

Keeping Track

Keeping track of what you've already counted is a problem for some contractors. If you are doing a take-off on a large set of plans-say, a shopping mall—it can be tedious work. The last thing you need to have happen is to lose your place or forget what you've already counted. To avoid this problem, mark each item on the plans as you count it. Using various colored pencils or highlighters is an easy way to keep track of what you have or have not counted. You might use a red pencil to check off each door and a blue one for each window. You might use a yellow highlighter on concrete work or a green highlighter on underground storm or sanitary lines.

Build in Contingencies

You should build a contingency in your take-off. If you think you are going to need 100 sheets of plywood, add a little to your count. How much you add will depend on the

Materials Take-off List.

Item name or use of piece	No. of pieces	Unit	Length in place	Size	Length	No. per length	Quantity
1. Footers	45	Pc	1'5"	2×6	10'	7	7
2. Spreaders	30	Pc	1'4"	2×6	8'	6	5
3. Foundation post	15	Pc	3'0"	6×6	12'	4	4
4. Scabs	20	Pc	1'0"	1×6	8'	8	3
5. Girders	36	Pc	10'0"	2×6	10'	1	36
6. Joists	46	Pc	10'0"	2×6	10'	1	46
7. Joist splices	21	Pc	2'0"	1×6	8'	4	6
8. Block bridging	40	Pc	1'10⅜"	2×6	8'	4	10
9. Closers	12	Pc	10'0"	1×8	10'	1	12
10. Flooring	800	BF	RL	1×6	RL	—	—

FIGURE 11-1   A sample materials take-off list.

size of the job you are figuring. A lot of estimators figure a contingency of 3 to 5 percent. Some contractors add 10 percent to their figures. Unless the job is small, I think a 10-percent add-on can cause you to lose the bid.

And don't forget to add something for waste. Rarely will you use every bit of your framing lumber, plywood, or sheet rock. You'll need to buy a certain quantity, but some will get thrown out in the trash. With accurate take-offs a small percentage for oversights will be sufficient. If you always seem to get on the job and run short of materials, a higher contingency will be needed.

Job Take-Off Form

Job Name: _____

Job Address: _____

Item	Quantity	Description
2" pipe	100'	PVC
4" pipe	40'	PVC
4" clean-out w/plug	1	PVC
2" quarter-bend	4	PVC
2" coupling	3	PVC
4" eighth-bend	2	PVC
Glue	1 quart	PVC
Cleaner	1 quart	PVC
Primer	1 quart	PVC

FIGURE 11-2   A sample job take-off form.

Keep Records

Keep records of your material needs on each job. Don't throw away your take-off sheets. When the job is done, compare the material actually used with what you estimated in your take-off. This will not only help you to see where your money is going; it will make you a better estimator. By tracking your jobs and comparing final counts with original estimates, you can refine your bidding techniques and win more jobs.

Pricing

Pricing your services and materials is an essential part of running a profitable business. If your prices are too low, you may be very busy, but your profits will suffer. If your prices are too high, you will be sitting around, staring at the ceiling, and hoping for the phone to ring. Somewhere between too low and too high is the optimum price for your products and services. The trick is finding out where that level is.

You must learn how to make your prices competitive without giving away too much of your profit. What's the right price? You can't pull your prices out of thin air. You must establish your pricing structure with research—lots of research. Talking to real-estate brokers and appraisers is an effective way to establish the market value of homes you plan to build. While brokers can be very helpful, I would spend most of my time talking to appraisers.

When I develop a new house plan that I'm interested in building, I meet with a licensed appraiser to get an opinion of value. Since I am a licensed broker, I don't have to talk with real-estate agents. I have access to multiple-listing services, which keep me apprised of what's on the market and what's been sold in recent months. By reviewing comparable sales and getting information from appraisers, I can target what the most realistic price will be for a home that I plan to build.

Some newcomers to the contracting business make a serious mistake. When they learn what their competitors are charging, they price their services far below the crowd, hoping to attract some business. However, setting extremely low prices can be the same as setting a trap for yourself; you may start a price war with the area's established competitors, driving prices and profit margins even lower, and you may alienate prospective customers. As a new company with low overhead, it's okay to price your product lower than the well-established companies, but don't price yourself into a deep hole.

When determining attractive prices, you must look below the surface. There are many factors that control what you are able to earn. Let's take a look at what is considered a profitable markup.

PRO POINTER

You must learn how to make your prices competitive without giving away too much of your profit. What's the right price? You can't pull your prices out of thin air. You must establish your pricing structure with research—lots of research. Talking to real-estate brokers and appraisers is an effective way to establish the market value of homes you plan to build.

What Is a Profitable Markup?

What is a profitable markup on materials? This can be a hard question to answer. It is not difficult to project what a reasonable markup is, but defining a profitable markup is not so easy.

Some contractors feel that a 10-percent markup is adequate. Others try to add 35 percent onto the price of their materials. Which group is right? You can't make that decision with the limited information I have given you. The contractors who charge a 10-percent markup may be doing fine, especially if they deal in big jobs and large quantities of materials. The 35-percent group may be justified in their markup, especially if they are selling small quantities of lower-priced materials.

Markup is a relative concept. Ten percent of $100,000 is much more than 35 percent of $100. Markups will need to be adjusted to meet the changes in market conditions and individual job

PRO POINTER

If you are selling everyday items that anyone can purchase from a local hardware or building-supply company, you must be careful not to inflate your prices too much. Customers expect you to mark up your materials, but they don't want to be gouged. If you installed light bulbs in that new light fixture and charged twice as much for the bulbs as the customer could have paid for them in the store, that customer is not going to be too happy.

requirements. You can, however, pick percentage numbers for most of your average sales. If you were in a repair business and were typically selling materials that cost around $20, a 35-percent markup would be fine if the market will bear it. If you are building and selling houses, a 10-percent markup on materials should be sufficient. To some extent, you have to test the market conditions to determine what price consumers are willing to pay for your materials.

If you are selling everyday items that anyone can purchase from a local hardware or building-supply company, you must be careful not to inflate your prices too much. Customers expect you to mark up your materials, but they don't want to be gouged. If you installed light bulbs in that new light fixture and charged twice as much for the bulbs as the customer could have paid for them in the store, that customer is not going to be too happy even though the amount of money involved in the light-bulb transaction is small; the principle of being charged double for a common item still rankles.

If you typically install specialty items, you can increase your markup. People will not be as upset when asked to pay a high markup on specialty items. A markup of 20 percent will almost always be acceptable on small residential jobs. When you decide to go above the 20-percent point, do so slowly and while testing the response of your customers. Realistically, you will probably find that a 10-percent markup will work best. You might be able to increase it to 15 percent, but that's usually about the limit for big-ticket items.

How Can Your Competitors Offer Such Low Prices?

This is a question almost all contractors ask themselves. There always seems to be some company out there that has a knack for winning bids and beating the competition. You can't help but wonder how that company does it.

Low prices can keep companies busy, but that doesn't mean the low-priced companies are making a profit. Gross sales are important, but net profits are what business is all about. If a company is not making a profit, there is not much sense in operating the business.

Companies that work with low prices fall into several categories. Some companies work on a volume basis. By generating a high volume, the company can make less money on each job and still make a decent profit by doing many jobs. This type of company is hard to beat because of its high volume.

When I was building in Virginia, I operated on a volume basis. My profit from a house was only about $7,000, but I was building as many as 60 of them a year. Keep in mind, this was back around 1984. If you do the math, you'll see that I was making a good living. The volume principle worked well for me in Virginia, but it won't work for me in Maine. I probably built more houses in a year while working in Virginia than are built by all the builders in my area of Maine in any given year. There simply isn't enough demand for housing in Maine to allow the volume approach to work very well.

Some companies sell at low prices because they are not really aware of their overhead and how much it really costs them to do business. For example, a carpenter who is making $16 an hour at a job might think that going into business independently and charging $25 an hour would be great. The carpenter might even start by charging only $20 an hour. From the carpenter's perspective, he is making at least $4 an hour more than he was at his job. Is he really making $4 an hour more? Yes and no; he is being paid an extra $4 an hour, but he is not going to get to keep much of it.

When this carpenter with his low hourly rate starts up in business, he may take a lot of work away from established contractors. Experienced contractors know they can't make ends meet by charging such low hourly rates. But our carpenter in this example, inexperienced as a business owner, is not

PRO POINTER

Companies that work with low prices fall into several categories. Some companies work on a volume basis. By generating a high volume, the company can make less money on each job and still make a decent profit by doing many jobs. This type of company is hard to beat because of its high volume.

aware of all the expenses he is incurring, and once the overhead expenses start eating away at what he thought was a great profit, he may be in for a real shock when he finds out that he is actually working for nothing.

A businessperson will soon learn that overhead expenses are a force to be reckoned with. Insurance, advertising, warranty and callbacks, self-employment taxes, and a lot of other expenses not initially considered will erode any profits quickly. And this type of businessperson will soon be out of business.

For established contractors competing against newcomers, being able to endure the momentary drop in sales will be enough to weather the storm. In a few months the new business will either be gone or have increased its prices to a more realistic level.

Pricing Services for Success and Longevity

Pricing services and materials correctly is directly related to your success and longevity. It can be very difficult to determine what the right price is for your time and materials. There are books with formulas and theories about how to set your prices, but these guides are not always on the mark. Every location can dictate variations in the prices the public is willing to pay. There are several methods that you can use to find the best fees for your business to charge; let's look at some of them.

Pricing Guides

Pricing guides can be a big help to the business owner with little knowledge of how to establish the value of labor and materials. However, these guides can cause you frustration and lost business. Most estimating guides provide a formula for adjusting the recommended prices for various regions. For example, a two-car garage that is worth $7,500 in Maine may be worth $10,000 in Virginia. These prices are purely hypothetical. The formula used to make this type of adjustment is usually a number that is multiplied against the recommended price. By using the multiplication factor, a price can be derived for services and materials in any major city.

PRO POINTER

Pricing services and materials correctly is directly related to your success and longevity. It can be very difficult to determine what the right price is for your time and materials. There are books with formulas and theories about how to set your prices, but these guides are not always on the mark.

The idea behind these estimating guides is a good one, but they are only guides. I have read and used many of these pricing books. From my personal experience, the books have not been accurate for the type of work I was doing. There are many times when the books are accurate, but I have never been comfortable depending solely on a general pricing guide.

PRO POINTER

Estimating books are very helpful, and while I don't use pricing figures from these books as my only means of setting a price, I do use them to compare my figures and to ensure that I haven't forgotten items or phases of work. Many bookstores carry some form of estimating guides in their inventory, and these types of books can be purchased on the Internet.

Estimating books are very helpful, and while I don't use pricing figures from these books as my only means of setting a price, I do use them to compare my figures and to ensure that I haven't forgotten items or phases of work. Many bookstores carry some form of estimating guides in their inventory, and these types of books can be purchased on the Internet.

Research

Research is an effective way to determine your pricing. When you look at historical data, you can find many answers to your questions. You can see how the economy swings up and down between supply and demand, between a buyer's market and a seller's market. You can see how prices have fluctuated up and down over the years, and you may be able to project the curve of the future. Historical data can be found by checking old newspaper ads on file in your local library, researching tax evaluations on homes, and talking with real-estate appraisers. Real-estate appraisers are an excellent source for pricing information. Most appraisers are willing to consult with contractors on an hourly basis.

Let's say that you are a general contractor and that you build houses, garages, additions, decks, and related home improvements. One way for you to establish the value of these services is to consult with a licensed real-estate appraiser. The appraiser can tell you the value the service will have on an official appraisal of the property. This doesn't guarantee that the values given by the appraiser are the best prices for you to use, but they are an excellent reference. By spending a little money to talk with an appraiser, you can save thousands of dollars in lost income. If your business involves providing goods and services for homes and businesses, appraisers are one of your best sources of pricing information.

Assume that the appraiser gives you a value of $2,000 for a 10-foot-by-10-foot deck. You could ask the appraiser to provide a written statement of value. Of course, the value will be general in nature and may not apply to all conditions, but the $2000 figure will be a solid average. You can use this written statement as a sales tool when you sit down with customers. If you are willing and able to sell the deck for $1,800, you can show the customer that it is lower than the average retail value of the deck.

You can also use the appraiser's figure in conjunction with numbers given in pricing guides and your own estimate to arrive at an average cost. In the case of new houses, the appraised value will typically be the selling price of your house. Unless your buyer is willing to put up cash for any overage, you will be limited to the appraised value.

Combined Methods

Combined methods are best when establishing your pricing principles. Use as much research and as many resources as possible to set your fees. Once you feel comfortable with your rates, test them. Ask customers to tell you their feelings. There is no feedback better than that from the people you serve.

Proper Presentation

Proper presentation is critical for business success. Even if your price is higher than the competition, you can still win the job with an effective presentation. There are many occasions when the low bidder does not get the job. As a contractor, you can set yourself apart from the crowd by using certain presentation methods. There are numerous ways to achieve an edge over the competition. The ways to win the bid battle can include the way you dress, what you drive, your organizational skills, and much more.

When you learn to convince a customer why you are worth more, you will make the most of your time and effort.

PRO POINTER

Proper presentation is critical for business success. Even if your price is higher than the competition, you can still win the job with an effective presentation. There are many occasions when the low bidder does not get the job. As a contractor, you can set yourself apart from the crowd by using certain presentation methods. There are numerous ways to achieve an edge over the competition. The ways to win the bid battle can include the way you drive, your organizational skills, and much more.

People are often willing to pay a higher price to get what they want, and it will be your job to convince the customer that you can provide better products and services. Let's look at some methods that have proven effective over the years.

Mailing Follow-Ups

You might be surprised by how many contractors mail estimates to customers and wait to be contacted; many never hear from the customer again. Mailing bids to potential customers may not be the most effective means. If customers are soliciting bids from your competitors as well, they may only look at the bottom line when making a decision. You want to make customers aware of the added benefits you can supply, so you need a better method of presentation than a cold mailing.

If you must mail your proposal, make sure you prepare a professional package. Use printed forms and stationery, not regular paper with your company name rubber-stamped onto it.

PRO POINTER

You might be surprised by how many contractors mail estimates to customers and wait to be contacted; many never hear from the customer again. Mailing bids to potential customers may not be the most effective means. If customers are soliciting bids from your competitors as well, they may only look at the bottom line when making a decision. You want to make customers aware of the added benefits you can supply, so you need a better method of presentation than a cold mailing.

Use a heavyweight paper and professional colors. Type your estimates and avoid using obvious correction methods to hide your typing errors. If you are mailing your price to the customer, you must make your presentation neat, well organized, attractive, professional, and convincing. With your computer, creating a neat presentation is easy— you can develop your own customized letterhead, add color if you have a color printer, and slip it into a nice binder that you can buy from your local office-supply company, and you've got a first-rate proposal.

Phone Facts

A formal proposal by phone is not the way to present your bid. When you call in your price, people can't visualize what they are getting or the person giving them this information. They can't linger over the estimate and evaluate it. Chances are good that the customer may write down your price and then lose the piece of paper it was written on. Phone estimates also tend to make contractors look indifferent, since they don't even take the time to present a professional, written proposal. In general, use the phone to get leads, set appointments, and follow up on estimates, but don't use it to give prices and proposals to possible customers.

Dress for Success

The dress code for contractors covers a wide range, depending on what you are selling and whom you are selling to. Whatever you wear, wear it well. Be neat and clean. Dress in a manner that you can be comfortable with. If you are uncomfortable in a three-piece suit, you will not project well to the customer. If you usually wear a uniform, wear it when presenting your proposal, but make sure it is clean and pressed. Jeans are acceptable and so are boots, but, again, both must be clean and neat. Avoid wearing tattered and stained clothing; it not only reflects on you but it also reflects on the product you wish to sell. You don't want the customer to be afraid to ask you to sit on the furniture.

PRO POINTER

The dress code for contractors covers a wide range, depending on what you are selling and whom you are selling to. Whatever you wear, wear it well. Be neat and clean. Dress in a manner that you can be comfortable with.

When you are deciding what to wear, consider the type of customer you will be meeting. If the customer is likely to be dressed casually, then you should dress casually. If you suspect a suit will fit in with your customer's attire, consider wearing a suit. Choose a wardrobe that will blend in with the customer. By overdressing, you might make the customer uneasy. And avoid flashy jewelry—that is usually a turn-off.

Your Vehicle

What you drive says a lot about you. If you drive an expensive car or truck, you are sending signals to the customer. When customers look out the window and see the contractor getting out of a fancy car, they think your prices will also be fancy. If you drive an old, dilapidated truck, they may assume that you are not very successful.

Choosing the best vehicle for your sales calls is a lot like choosing the proper clothing. You want a vehicle that will make the right statement. A clean van or pickup is fine for most any occasion. Cars are okay for sales calls, but avoid pulling up to your customer's house in an expensive luxury model unless you are catering to high-end clients.

Confidence, the Key to Success

Confidence is the key to success. You must be confident in yourself, and you must create confidence in the mind of the consumer. If you can get the customer's confidence, you can almost always get the sale. You will gain your own confidence through experience, but you must learn how to build confidence with your customers.

Gaining the confidence of your customers can often be done by the way in which you talk to them. If you are able to sit down with customers and talk for an hour, the chances of getting the job increase greatly. Showing customers examples of your work can build confidence. Letters of reference from past customers will help in establishing trust. If you have the right personality and sales skills, you can create confidence by simply talking.

As a business owner you are also a sales professional, or at least you need to learn how to become one. Unless you hire outside sales staff, you are the person customers will be dealing with. If you learn basic sales skills, you will become a successful contractor even if your prices are higher than your competitors.

Know Your Competition

You must know your competition. How you price your services and materials will be affected by the competition. Your prices should be in the same range as your reputable competitors. If your prices are too high, you won't get much work. If your prices are too low, you may get lots of work and low profits.

Effective Estimating Techniques

Effective estimating techniques come in many forms. What works for one person doesn't work for another. By learning from your mistakes and successes, you can develop your own profitable estimating methods and use them over and over again. You may have to change your technique from customer to customer, but the same basic principles that work for you in one sale will work in another.

You can learn technical aspects of estimating by referring to estimating and pricing guides. Look at the information in your past projects and dig out unit costs and square-foot costs. Ask your subcontractors how much they would charge for certain jobs.

Much of what you learn will come from experience. Learning from your mistakes can be expensive, but you don't soon forget your costly lessons. One of the most important factors in effective estimating is organization. When you are well organized, you are more likely to complete a thorough estimate.

PRO POINTER

You can learn technical aspects of estimating by referring to estimating and pricing guides. Look at the information in your past projects and dig out unit costs and square-foot costs. Ask your subcontractors how much they would charge for certain jobs.

Cost Projections

Item/Phase	Labor	Material	Total
Plans			
Specifications			
Permits			
Trash container deposit			
Trash container delivery			
Demolition			
Dump fees			
Rough plumbing			
Rough electrical			
Rough heating/ac			
Subfloor			
Insulation			
Drywall			
Ceramic tile			
Linen closet			
Baseboard trim			
Window trim			
Door trim			
Paint/wallpaper			
Underlayment			
Finish floor covering			
Linen closet shelves			
Closet door & hardware			
Main door hardware			
Wall cabinets			
Base cabinets			
Countertops			
Plumbing fixtures			
Trim plumbing material			
Final plumbing			
Shower enclosure			
Subtotal			

FIGURE 11-3   A sample form for calculating cost projections. *(continued)*

Item/Phase	Labor	Material	Total
Light fixtures			
Trim electrical material			
Final electrical			
Trim heating/ac material			
Final heating/ac			
Bathroom accessories			
Clean up			
Trash container removal			
Window treatments			
Personal touches			
Financing expenses			
Miscellaneous expenses			
Unexpected expenses			
Margin of error			
Subtotal from first page			
Total estimated expense			

FIGURE 11-3   A sample form for calculating cost projections.

Using Sustainable Building Practices to Make More Money

Is using sustainable building practices to make more money a bad procedure? Of course it isn't. You are in business to make money. By using environmentally friendly methods in building, you are contributing to the longevity of the world as we know it. Many people will pay more to preserve the environment. So I ask again, is this a bad thing? No, it is not.

People need places to live and work. Construction will take place. By using sustainable products and practices, you are being responsible. It is well known that it cost more to build green, but many people are willing to pay the price. This is how you can make more money doing the right thing.

What's the catch? There isn't one. It is a win-win situation for all involved. You make more money, customers are happy, and the environment benefits from your efforts. How does it get any better than that? Have I got your attention? Okay, let's move on in our exploration of how you can cash in on the booming green building business.

Why Pay More?

Why pay more for the same house or office building? This is a question that some builders ask themselves when they consider becoming green builders. Not everyone is willing to pay more for environmentally friendly construction. There is, and probably always will be, a percentage of people who will seek the cheapest building costs that they can find.

Would you pay more to save a tree or to reduce the demand for fuel to heat your home? Is it an investment in the future, or is it a gouge in your pocket today? Everyone in the market has to answer these questions. Most of us understand and respect the need to protect our environment for wildlife and future generations. But are we willing to pay the price for doing the right thing? This is up for debate, but there is a growing trend in responsible building practices.

Consider our appreciation of vehicles. There are those who have electric cars and those who have what is often called a gas-guzzler. Not many builders would be happy working out of a compact car. They need trucks and vans. What does this have to do with sustainable building? It opens a door to options. You don't have to go all green to do some good for your environment.

Doctors often say that moderation is the key to good health. This can also be true in sustainable building. It is not always practical to build all green. What I have just stated will not go over well with some people, but it is true. Change takes time. It is coming; you can count on it. But how soon will it come to your town or city?

This book is meant to help you become a green builder. Having been in construction for about 30 years, I know the ropes from the old days and have been learning them more and more each year. My way is personal research and education. I believe that overcoming complacency is the first step in convincing your customers. That said, let's see how you can profit from going green.

What will make people pay more per square foot for a similar, but green, house? Education will. You have to educate your customers. Putting up a sign that says you are a green builder will not get it done. The process of going green will take effort on your part if it is to work successfully.

Plenty of people are already looking for environmentally friendly building projects. These potential customers have strong feelings for the environment and are willing to pay more to preserve it. Since builders normally

PRO POINTER

Plenty of people are already looking for environmentally friendly building projects. These potential customers have strong feelings for the environment and are willing to pay more to preserve it. Since builders normally make their money as a percentage of the total cost to build a building, the end result is more money in your pocket. But you have to become known as a green builder to take advantage of this.

make their money as a percentage of the total cost to build a building, the end result is more money in your pocket. But you have to become known as a green builder to take advantage of this.

Builders who establish themselves as sustainable builders are sought after every day. Once you are educated, you can educate your potential customers. Advertising will be a factor, but your depth and breadth of knowledge will be the key. Once you have a few projects under your belt, you should be off and running.

Sales Training

Consider taking some sales training to increase your sales and profits. As a sustainable builder, you represent a good cause, but the expense of being responsible can be difficult to sell. If you are not comfortable with selling, think about hiring someone who is. You need to get your message, your goal, and your values in front of potential customers. If your time is better spent in the field as a supervisor, hire a salesperson.

Ideally, you should strive to be the best salesperson you can be. After all, it is your business, your life, and your money. Learning to sell yourself and your services doesn't have to take a long time. You can learn a lot by reading books that explain how to sell. Some people think that salespeople are less than reputable. This does not have to be the case. Think of your sales pitch as an educational lesson.

PRO POINTER

Consider taking some sales training to increase your sales and profits. As a sustainable builder, you represent a good cause, but the expense of being responsible can be difficult to sell. If you are not comfortable with selling, think about hiring someone who is. You need to get your message, your goal, and your values in front of potential customers.

Senseless waste of resources is not something to be proud of. On the other hand, building responsibly is providing a solid service for the right reasons. Many customers respect this. As you become a noted green builder, you will begin to build a following of loyal customers. News of bad contractors travels quickly. Getting out the message about good contractors takes longer, but it is a much better position to be in. Build what you believe in. People are willing to pay more for responsible building when they understand what they are doing. Selling takes many forms. In this case, your sales pitch will be as simple as making people aware of what the additional costs are for and how your products maintain the security of life as we know it.

Do the Math

Do the math to see how you will make more money as a green builder. It is common knowledge that builders use a percentage markup of the cost of construction to arrive at a sales price. The percentage varies, but a 20-percent markup is not uncommon. If you build a house that costs you $300,000 to build, it would not be unreasonable to price it at $360,000. This can give you a strong income on a per-house basis.

Sustainable materials cost more. This is no secret. When costs go up, the profit potential can go up. A traditional home that costs $300,000 to build will usually cost more to build as a green home. In doing this, you can base your percentage of income on the elevated price. The percentage might not be dollar for dollar, but there is room for increased income when you build responsibly.

Know Your Product

Before you present yourself as a green builder, know your product inside and out. People who are in the market for green homes have normally done plenty of research. Before you go green, you have to understand the concept and have answers to questions that you may be asked. Nothing will kill a deal quicker than a contractor who is not informed about the most current trends and options. If you run into potential customers who know more about your business than you do, it is highly unlikely that they will become your customers.

Many builders like to build. Not as many of them like to conduct research. Get over it. If you are going green, you are going to have to educate yourself before you can educate your buyers. Do your due diligence. Become immersed in the learning curve. This is the single largest favor you can do yourself in going green.

PRO POINTER

Before you present yourself as a green builder, know your product inside and out. People who are in the market for green homes have normally done plenty of research. Before you go green, you have to understand the concept and have answers to questions that you may be asked. Nothing will kill a deal quicker than a contractor who is not informed about the most current trends and options. If you run into potential customers who know more about your business than you do, it is highly unlikely that they will become your customers.

Preserving the Future

As a green builder, you are making your contribution to preserving the future. You can feel good about what you are doing and you should be able to see a higher profit margin in the process. It will not come easily. People need to be made aware of their options and what their decisions will cost.

Will a homebuyer give up the delicate chandelier in the dining room to save a tree? Would the buyer consider leaving the dining room out of the blueprints to pay for the additional cost of a green home? If you are a responsible builder, will there be responsible buyers? Not everyone is going to lose sleep over trees falling in the forest, but many buyers are tuned into being environmentally responsible. These are your customers. Once you learn how to perform properly as a sustainable builder, you will have a wealth of opportunity before you. This is your chance to help save the environment while cashing in on it yourself. It truly is a win-win situation, so go for it.

PRO POINTER

As a green builder, you are making your contribution to preserving the future. You can feel good about what you are doing and you should be able to see a higher profit margin in the process. It will not come easily. People need to be made aware of their options and what their decisions will cost.

Growing Your Business the Smart Way

How much is your public image worth? The type of business image you present can mean the difference between success and failure. People quickly learn to associate a logo with its owner. Logos can do a lot for you in all your display advertising by creating familiarity between readers and your company. The name you choose for your company is important and needs careful consideration. Some names are easier to remember than others. A name can conjure a mental image, an image you want for your business. If it is your intention, long range, to sell your business, the company name should not be too personalized; the new owner may not like owning a business with your name as part of the company name.

A company image can affect the type of customers the business attracts. Your business advantage can originate from your involvement in community organizations. How you shape your business image may set you apart from the competition.

It is difficult to put a price on your public image. While it is difficult to set a monetary value on your company image, it is easy to see how a bad image will hurt your business. Your public image has many facets. Your tools, trucks, signs, advertising, and uniforms will all have an impact on your corporate image. This chapter is going to detail how these and other factors work to make or break your business.

Public Perception Is Half the Battle

How the public perceives your business is half the battle. If you give your customers the impression of being a successful business, your chances of being successful will increase. On the other hand, if you don't develop a strong public image, your business may sink into obscurity.

How does the public judge your public image? There are many factors that contribute to how the public perceives your business. Take your truck as an example. What kind of image do you think an old, battered pickup, with bald tires and a license plate hanging from baling wire, will project? Would you rather do business with this individual or a person driving a late-model, clean van that has the company name professionally lettered on the side? Which truck points to the most company success and stability? Most people would prefer to do business with a company that gives the appearance of being financially sound. This doesn't mean you have to have flashy new trucks, but they should be well maintained.

It is important to have your company name on the business vehicles. The more often people see your trucks around town, the more they will remember your name and develop a sense of confidence. It is acceptable to use magnetic signs or professional lettering, but don't letter the truck with stick-on letters in a haphazard way. Remember, you are putting your company name out there for all the world to see. You can end up attracting attention, but not the kind you'd like to attract.

Designing your ad for the phone directory is another major step in creating a company image. As people flip through the pages of the directory, a handsome ad may stop them in their tracks. An eye-appealing ad can get you business that would otherwise be lost to competitors.

While we are talking about phone directories, let's not forget about phone manners. Telephones often pro-

PRO POINTER

It is important to have your company name on the business vehicles. The more often people see your trucks around town, the more they will remember your name and develop a sense of confidence. It is acceptable to use magnetic signs or professional lettering, but don't letter the truck with stick-on letters in a haphazard way. Remember, you are putting your company name out there for all the world to see. You can end up attracting attention, but not the kind you'd like to attract.

vide the first personal link between your business and potential customers. If you lose customers at the inquiry stage, your business will suffer. You could lose customers if you allow small children to answer your business phone. An answering machine with an abrupt message is a sure way to lose potential business. Answering machines may cost you some business, too, but they are an acceptable form of doing business. Customers are calling and they expect a professional response. If you put a tape in your answering machine that is non-professional or offensive, would-be customers are sure to hang up.

PRO POINTER

If your company image is strong, customers will come to you. They will see your trucks, job signs, and ads and call you. When a customer calls a contractor, he or she is usually serious about having work done, and your company image will help you in landing the job. By building and presenting the proper image, you are halfway toward making the sale.

Any professional salesperson will tell you that to be successful, you must always be in a selling mode. It doesn't matter where you are or what you're doing—you must be ready to cultivate sales and your image every day and in every way. You can't afford to let your business image slip.

If your company image is strong, customers will come to you. They will see your trucks, job signs, and ads and call you. When a customer calls a contractor, he or she is usually serious about having work done, and your company image will help you in landing the job. By building and presenting the proper image, you are halfway toward making the sale.

Picking a Company Name and Logo

Picking a company name and logo should be considered a major step in building your business image. The name and logo you choose will be with the company for many years. Before you decide on a name or a logo, you should do some research and some thinking. There will be many questions to ask yourself. For example, do you plan to sell the business in later years? If you do, pick a name that anyone could use comfortably. A name like Pioneer Green Design can be used by anyone, but a name like Ron's Green Remodeling is more difficult for a new owner to adopt. This is only one example; let's move on to other considerations in choosing a name and logo.

Company Names

Company names can say a lot about the business they represent. For example, High-Tech Heating Contractors might be a good name for a company specializing in new heating technologies and systems. Solar Systems Unlimited could be a good name for a company that deals with solar heating systems. Authentic Custom Capes would make a good name for a builder that specialized in building period-model Cape Cods. How would a name like Jim's Custom Homes do? It might be all right, but it doesn't say much. A better choice might be Jim's Affordable First-Time Homes. Now the name tells customers that Jim is there for them, with affordable first homes. See how a name can influence the perception of your business?

Your company name should be one that you like, but it should also work for you. If you can imply something about your business in the name, you have an automatic advantage. It has been proven that people remember things that they are shown repetitively. When you run an ad in the paper every week, people will remember your ad. Even though readers may not realize it, they are committing your company name or logo to their subconscious. Then, when these people scan the pages of a phone directory for a contractor and run across your name or logo, you have an edge.

Since these potential customers have been given a steady dose of your advertising, your name or logo will stick out from the crowd. Without thinking about it, people who have had regular exposure to your advertising will remember something about the ads.

Since advertising is expensive, it makes sense to get as much bang for your advertising buck as possible. If you were scanning through the newspaper and noticed a company name like Tanglewood Enterprises, what would you associate the name with? That name is not descriptive and could be used for any number of different types of businesses.

> **PRO POINTER**
>
> Company names can say a lot about the business they represent. For example, High-Tech Heating Contractors might be a good name for a company specializing in new heating technologies and systems. Solar Systems Unlimited could be a good name for a company that deals with solar heating systems. Authentic Custom Capes would make a good name for a builder that specialized in building period-model Cape Cods. How would a name like Jim's Custom Homes do? It might be all right, but it doesn't say much. A better choice might be Jim's Affordable First-Time Homes. Now the name tells customers that Jim is there for them, with affordable first homes.

On the other hand, if you saw a name like Deck Masters, Inc., you would associate the name with decks. If the name was White Lightning Electrical Services, you would think of an electrical company. A name like Homestead Homes gives a clear impression of a company that offers warm, comfortable housing.

The more you can equate the name of your business to the type of business you are in, the better off you will be. If you can add descriptive words, your customers will know more about your business just from the name.

It also helps when choosing a name for your company to find words that flow together smoothly. How does Pioneer Plumbing sound to you? Both words start with a "P," and the words work well together. A name like Ron's Remodeling sounds good, and so does a name like Mike's Masonry. In contrast, a name like Englewood Heating and Air Conditioning is not bad. A name like Septic Suckers flows well and might be fitting for a company offering service to septic tanks, but it may be offensive to some people. Your company name says a lot about your business. Maybe your company name would be something like Best Builders or Built to Perfection. Find a name that identifies what you do best.

Logos

Logos can be as important as your company name. Logos, the symbols that companies adopt to represent them graphically, play an important role in marketing and advertising. While people may not remember a specific ad or even a company name, they are likely to remember distinctive logos. If you put your mind to it, I'll bet you can come up with at least ten logos that stick in your mind.

Major corporations know the marketing value of logos and invest considerable time and money in coming up with just the right symbol to represent their corporation. Like slogans and jingles, logos are often much easier to remember than company names.

Do you remember which gas company used to put a tiger in your tank? In the cola wars, who has the "right thing"? If you were shopping for tires, whose ad would you remember? My guess is that

PRO POINTER

Logos can be as important as your company name. Logos, the symbols that companies adopt to represent them graphically, play an important role in marketing and advertising. While people may not remember a specific ad or even a company name, they are likely to remember distinctive logos. If you put your mind to it, I'll bet you can come up with at least ten logos that stick in your mind.

you might remember the baby riding around in a certain brand of tire. Who lets you reach out and touch someone—by phone, of course? You see, it is easy to remember ads, slogans, and jingles. It is equally easy to remember logos.

Your logo doesn't have to be complex. In fact, it might be nothing more than the initials of your company name. Then again, you may have a very complex logo, one that incorporates an image of what your business does. One of my favorite logos was used by a real-estate company. The logo was a depiction of Noah's ark, complete with animals. In the ad featuring the ark were the words "Looking for Land?" The business was selling land, and the ark logo was humorous and fitting for the occasion.

PRO POINTER

If it is difficult for you to create images and marketing, it may serve you well to consult with a specialist in the field. Choosing the proper name and logo is important enough to warrant investing some time and money. Of course, you know your financial limitations, but if you can afford it, get some professional advice in designing the image of your company.

If it is difficult for you to create images and marketing, it may serve you well to consult with a specialist in the field. Choosing the proper name and logo is important enough to warrant investing some time and money. Of course, you know your financial limitations, but if you can afford it, get some professional advice in designing the image of your company.

How Your Image Affects Your Clientele and Fee Schedule

How does your image affect your clientele and fee schedule? Image may not be everything in business, but it is a big part of your success. People have become wary of contractors. The public has read all the horror stories of ripoffs and contractor con artists, some of which are true, and the public does have a right to be concerned. With the growing awareness of consumers, image is more important than ever before. Let's take a quick look at three examples of making a sales call.

The Visual Image

The visual image that you and your business project can influence the profits of your company. In our first example, a contractor goes on an estimate wearing well-worn work clothes and driving a truck that has seen much better days. While this image may not offend or alienate some customers, it will surely turn many customers off.

On the other hand, you can go overboard image-wise. You drive up in a luxury car wearing a suit that cost more than the first contractor's truck. Some homeowners will relate and respond well to that image, especially if you are catering to the high end of the market. So you need to moderate and adjust the image you are trying to project to the type of project you are selling and the type of customer you are selling it to.

> **PRO POINTER**
>
> If you dress neatly, even in casual clothing, and drive a respectable, well-maintained vehicle, your odds of appealing to the masses improve. By wearing clothes that make you believable as a skilled tradesperson, you give the impression of someone who knows the contracting business. Your vehicle looks professional and successful. For me, this combination has always worked best.

If you dress neatly, even in casual clothing, and drive a respectable, well-maintained vehicle, your odds of appealing to the masses improve. By wearing clothes that make you believable as a skilled tradesperson, you give the impression of someone who knows the contracting business. Your vehicle looks professional and successful. For me, this combination has always worked best.

The same basic principles apply to your office. If your office is little more than a hole in the wall with an answering machine, an old desk, and two broken chairs, people will be concerned about the financial stability of your business. However, if your office is staffed with several people, decorated in expensive art and furnishings, and in an expensive location, customers will assume your prices are too high. It generally works best to hit a happy medium with your office arrangements.

Fee Factors

What does image have to do with the fee you charge for your services? To a large extent, people feel that they get what they pay for. The image you present has a direct effect on your fees.

If you convince potential customers that you are a professional, the customers will be willing to pay professional fees. Extend your image by making the customers feel secure doing business with you, and you have leverage for even higher fees. When you become a specialist, you may be able to demand higher fees. Think about it: Who gets a higher hourly rate, your family doctor or a heart specialist? When you convince the customer that nobody builds a better green house than you do, you are building a case for higher fees.

For years I specialized in kitchen and bath remodeling. My crews did nothing but kitchen and bathroom remodeling. When you do the same type of work day in and day out, you get pretty good at it.

With my experience in this specialized field of remodeling, I could anticipate problems and find solutions before most of my competitors could. This specialized experience made me more valuable to consumers. I could snake a two-inch vent pipe up the wall from their kitchen to their attic without cutting the wall open. I could predict with accuracy how long it would take to break up and patch the concrete floor for a basement bath. In general, I became known as a competent professional in a specialized field, and I could name my own price, within reason, for my services. You can do the same thing with green homebuilding.

If you have a special skill and can show the consumer why you are more valuable than your competitor, the consumer is very likely to pay a little extra for your expertise. Building a solid image as a professional that specializes in a certain field has its advantages.

Once You Cast an Image, It Is Difficult to Change

Once an image is cast, it is difficult to change. If you are already operating an established business, it is more difficult to change your image than it is to create a new one. If during the building of your company image you have found flaws, work to change them. With enough time, effort, and money, you can make a difference in your company image.

Let's say you started your business without much thought. You picked a name out of thin air, and you never got around to designing a logo. Now you realize that you have hurt the prospects

PRO POINTER

Once an image is cast, it is difficult to change. If you are already operating an established business, it is more difficult to change your image than it is to create a new one. If during the building of your company image you have found flaws, work to change them. With enough time, effort, and money, you can make a difference in your company image.

of your business getting off the ground. What should you do? You must make changes to correct your mistakes.

You can create a logo, and you can change the direction of your company, but changing the name can get tricky. If you change the name abruptly, you may lose existing customers. How will you accomplish your goal of changing the name? The procedure is not as difficult as you may think.

When you want to change the name of your existing company, plan a direct-mail campaign to your existing customer base. Send out letters to all of your customers, advising them of your new company name. Explain that due to growth and expansion, for example, you are changing the name of the company to reflect your new services. Impress upon the existing customers that the

PRO POINTER

When you want to change the name of your existing company, plan a direct-mail campaign to your existing customer base. Send out letters to all of your customers, advising them of your new company name. Explain that due to growth and expansion, for example, you are changing the name of the company to reflect your new services. Impress upon the existing customers that the company has not been sold and is not under new management—if you have a bad image to overcome, however, the new-management announcement might be a good idea.

company has not been sold and is not under new management—if you have a bad image to overcome, however, the new-management announcement might be a good idea.

Start running new advertisements with the new company name and logo. Build new business under your new name and convince past customers to follow you in your expansion efforts. By taking this approach, you get a new public image without losing the bulk of your past customers. While this approach will work, it is better to take the time and effort to create a good image when you begin the business. It is always easier to do the job right the first time than it is to go back and correct mistakes.

Set Yourself Apart from the Crowd

In order to make your business better than average, you must set yourself apart from the crowd. You can do this with a logo, company colors, slogans, and any other ways that are appropriate for your business. We have already talked about logos, so let's look at some other ways to give your company a unique identity.

Company Colors

Company colors are one way to attract attention and become known all over town. If you don't think color makes a difference, ask the cab drivers who ride around in yellow cars. Need another example? How about the colors red, white, and blue—what do they mean to you? Colors can have a strong impact on what we think of and the context in which we think of it.

There are consultants who specialize in colors. These professionals work with companies to design colors to influence consumers. Different colors affect how people think, their mood, and how they react. We've all heard about bulls charging a red flag; would they charge a green flag? You see, whether the color of the flag makes a difference is not as important as our mental image of what the color says. Since we are taught that red will make the bull mad, we tend to believe it, but I'll bet there are some bulls that would be just as happy charging a blue flag.

Look at how our culture has classified the personalities of people based upon hair color. People with red hair are said to have short fuses and high tempers. Blondes are supposed to have more fun. Obviously, the color of a person's hair doesn't make them dumb, fun, or hot-tempered.

Choosing the right colors for your company is important. How seriously would you take a builder who pulled up in a pink van with flowers painted on it? The color and decoration of the van may have no bearing on the technical ability of the builder, but it does create an immediate impression. You should choose your company colors with care.

The color of your trucks may be dictated by the color of the truck you presently own. It is more impressive to see a fleet of trucks that are uniform in color and design than it is to see a parade of trucks that include various makes and colors. A unified fleet gives a better impression.

Colors are also important in your business stationery. It would be inappropriate to use fluorescent orange for your letters and lime green for your envelopes. Certainly,

PRO POINTER

Company colors are one way to attract attention and become known all over town. If you don't think color makes a difference, ask the cab drivers who ride around in yellow cars. Need another example? How about the colors red, white, and blue—what do they mean to you? Colors can have a strong impact on what we think of and the context in which we think of it.

these colors would attract attention and be remembered, but the impression would not likely be the one you wanted to create. For most businesses tan, ivory, light blue, or off-white are acceptable stationery colors.

Color is also important in your truck lettering and job-site signs. If your truck is dark blue, white letters will show up better than black letters. If the truck is white, black or blue letters would be fine. For job signs, it is important to pick a background and a letter color that contrast well. You want the sign to be easy to read from a distance, so size is also important. When you talk with your sign painter or dealer, you can review samples of how different colors work together. Now, let's examine the value of company slogans.

Slogans

Slogans are often remembered when company names are not. If you will be advertising on radio or television, slogans are especially important. Since radio and television provide audible advertising, a catchy slogan can make its mark and be remembered. When advertising in newspapers or other print ads, slogans show the readers key words to associate with your company.

Try a simple test. When I ask you the question in the next sentence (no fair peeking), think of one company as fast as you can. Which pizza company

> **PRO POINTER**
>
> Slogans are often remembered when company names are not. If you will be advertising on radio or television, slogans are especially important. Since radio and television provide audible advertising, a catchy slogan can make its mark and be remembered. When advertising in newspapers or other print ads, slogans show the readers key words to associate with your company.

delivers? If you thought of Domino's, my guess was right. Domino's has been a great success in a highly competitive business. The logo on their box is a domino. A lot of pizza places deliver, but if you live in an area where Domino's is available, you can't think of pizza delivery without thinking of Domino's. This fact is no accident. I'm sure the brains behind Domino's have spent huge sums of money to develop this image.

The golden arches are another example of good marketing, as is the slogan that lets you have it your way. All major food franchises have established logos, slogans, colors, and more. Their marketing and advertising are expensive, but they work. Advertising only costs a lot of money when it doesn't work. If expensive advertising works, it makes you money.

Thinking of a slogan for your business might take a while, but it's worth the effort. If you need inspiration, look around you at other successful companies. Examine their slogans to gain ideas for yours, but never use someone else's slogan.

Build Demand for Your Services through a Strong Image

You can build demand for your services through a strong image. People like to deal with winners. If your company has the reputation of being fair, professional, competent, and dependable, customers will seek you out.

Building business demand through a strong image can be done in several ways. One way is to build your business a little each time you serve a customer. This will build a word-of-mouth referral system. Word-of-mouth referrals are the best business you can get. But if you don't want to wait for the results of customer recommendations, you can use advertising.

> **PRO POINTER**
>
> You can build demand for your services through a strong image. People like to deal with winners. If your company has the reputation of being fair, professional, competent, and dependable, customers will seek you out.

Advertising is a very powerful business tool. In skilled hands, advertising can produce fantastic sales results. Consider this: You are about to move to a new city; which real-estate brokerage will you call for relocation help? I would guess you would call a brokerage where all the brokers and agents wear gold coats. The gold-coat brokers get a lot of visibility on TV, radio, and in print ads. Once you hear a hundred times how they are the best real-estate team around, you might start to believe it.

You may not know anything about a particular brokerage, but advertising plants the seed that the gold coats mean success. If you buy into the advertising, you are likely to call these brokers. If the broker doesn't make a good personal impression, you may choose another brokerage, but at least you called the gold team. This same strategy can work for you.

When operating a contracting business, advertising alone will not get the job done. You or your company representatives will have to keep the ball rolling once you are in touch with potential customers. Talk to some professionals in the field of marketing, and I think you will be surprised at the results you can achieve.

Joining Clubs and Organizations to Generate Sales Leads

Joining clubs and organizations is an excellent way to generate sales leads. As a business owner, you must also be a salesperson. When you join local clubs and community organizations you meet people, and these people are all potential customers.

By becoming visible in your community, your business will have a better chance of survival. If you support local functions, children's sports teams, and the like, you become known. You can use the local opportunities to build your business image. When citizens see your company name on the uniforms of the local kids' baseball team, they remember you. Further, they respect you for supporting the children of the community. You can take this type of approach to almost any level. After you have established a public awareness of your business, you should get busy. You can't afford to fall back on your marketing and advertising needs.

PRO POINTER

Joining clubs and organizations is an excellent way to generate sales leads. As a business owner, you must also be a salesperson. When you join local clubs and community organizations you meet people, and these people are all potential customers.

Marketing and advertising may well be the most important lessons for new business owners to learn. While it is true that marketing and advertising alone will not make a business a success, they are critical elements in building a thriving business. If you don't do a good job with your marketing and advertising, you won't have a chance to show customers what you can do.

There Is No Business without Sales

There is no business without sales. To get the opportunity to generate sales, most businesses must advertise. Without advertising, the average business will have a hard time gaining customer interest. If no one knows your business exists, how will they contact you for service? Since public exposure is paramount to the success of your business, so is a strong marketing plan and effective advertising.

Too many contractors fail to see the importance of marketing and advertising. For some reason, many contractors think the public will seek them out. Let me repeat myself: If the public doesn't know you exist, they can't very well seek you out. Regardless of how good you are at what you do, you won't get much work without making people aware of your services.

To get busy and stay busy, you need regular sales. Marketing and advertising can provide you with sales leads. It will be up to you or your salespeople to convert the leads into closed sales, but you must start by getting prospects wanting what you have to offer. Advertising is the most effective way to generate leads quickly.

PRO POINTER

To get busy and stay busy, you need regular sales. Marketing and advertising can provide you with sales leads. It will be up to you or your salespeople to convert the leads into closed sales, but you must start by getting prospects wanting what you have to offer. Advertising is the most effective way to generate leads quickly.

Marketing Is a Pivotal Point for Any Business

Marketing is a pivotal point for any business. If you have the ability to perform a good market survey, you should be able to generate a vast amount of business. Advertising is the act of putting your message in front of consumers. You can advertise in newspapers, on radio, on television, by direct mail, or in many other ways. Marketing is not just advertising. Marketing is reading the business climate. When you track your advertising results, design your ads, develop sales strategies, and define your target market, you are exhibiting marketing skills.

Marketing is much more complicated than advertising. Advertising your business requires little more than the money needed to pay for your ads. Marketing demands an extension of your normal senses. You must be able to read between the lines and determine what the buying public wants. There are many books available on marketing. Professional seminars teach marketing techniques. Many community colleges offer courses in marketing. With enough

PRO POINTER

Marketing is much more complicated than advertising. Advertising your business requires little more than the money needed to pay for your ads. Marketing demands an extension of your normal senses. You must be able to read between the lines and determine what the buying public wants.

effort and self-study, you can become very efficient with your marketing ploys. If you want to have a business with a long life, you should expend the energy to develop effective marketing skills.

Should You Enlist Commissioned Salespeople?

This is a good question, and the answer depends on what your business goals are. Commissioned salespeople can make a dramatic difference in your business. Commissioned salespeople can generate a high volume of gross sales, and since you are paying the sales staff only for what they sell, an army of sales associates can be mighty enticing. However, a high volume of sales can create numerous problems. You may not have enough help to get the jobs done on time. You might have to buy new trucks and equipment and increase your workforce quickly without having enough time to determine if they are fully qualified. The increased business may tie you to the office and cause your field supervision to suffer. There are many angles to consider before bringing a high-powered sales staff online.

> **PRO POINTER**
>
> Commissioned salespeople can make a dramatic difference in your business. Commissioned salespeople can generate a high volume of gross sales, and since you are paying the sales staff only for what they sell, an army of sales associates can be mighty enticing. However, a high volume of sales can create numerous problems.

The Benefits of a Sales Staff

The benefits of a sales staff are many. If you find the right people to represent your company, you can enjoy increased sales. By having commissioned salespeople, you don't have the normal overhead of employees, since you only pay for what you get. Good salespeople will generate deals that may otherwise never come your way. A strong deal closer will make deals happen, so you have quick sales. Sales professionals can convert a simple job into a major job for you. With the right training and experience, sales professionals can get more money for a job than the average contractor would. It is clear that for some businesses a sales staff is a powerful advantage.

The Drawbacks to Commissioned Salespeople

The drawbacks to commissioned salespeople may outweigh the advantages. Some salespeople will tell the customer anything they want to hear to get a signature on the contract. As the business owner, you will have to deal with this form of sales embellishment at some point during the job. The customer might tell you that the salesperson assured them that they would get screens with their replacement windows when you had not figured screens into the cost of the job. The salesperson might have promised that the job could be done in two weeks when in reality the job will take four weeks. This type of sales hype can cause some serious problems for you and your workers.

Most sales associates are not tradespeople. They don't know all the technical aspects of a job; they only know how to sell the job, not how to do it. A salesperson might tell a prospect that putting a bathroom in the basement is no problem, when in fact such an installation requires a sewer pump that adds nearly $800 to the cost of an average basement bath. There are many times when an outside sales staff undersells a job. Sometimes they sell the job cheap to get a sale. At other times the wrong price is quoted out of ignorance. In either case, you, as the business owner, have to answer to the customer.

Getting too many sales too quickly can be as devastating as not having enough sales. If the salesperson you put in the field is good, you might be swamped with work. This can lead to problems in scheduling work, the quality of the work turned out, field supervision, cash flow, and a host of other potential business killers.

Sending the wrong person out to represent your company can have a detrimental affect on your company image. If the salesperson is dishonest or gives the customer a hard time, your business reputation will suffer. Deciding when to use commissioned salespeople is your decision. But let me tell you, don't make the decision lightly. There is no question that the right salespeople can make your business more profitable. However, there is also little doubt that the wrong sales staff can drive your business into the ground.

PRO POINTER

Getting too many sales too quickly can be as devastating as not having enough sales. If the salesperson you put in the field is good, you might be swamped with work. This can lead to problems in scheduling work, the quality of the work turned out, field supervision, cash flow, and a host of other potential business killers.

If you decide to use commissioned salespeople, I suggest you go with them on the first few sales calls. When you are interviewing people to represent your company, remember that they are sales professionals. These people will be selling you in the interview with the same tenacity that they will use on prospects in the field. Go into the relationship with your eyes wide open. Don't take anything for granted, and check the individuals out for integrity and professionalism.

PRO POINTER

If you decide to use commissioned salespeople, I suggest you go with them on the first few sales calls. When you are interviewing people to represent your company, remember that they are sales professionals. These people will be selling you in the interview with the same tenacity that they will use on prospects in the field. Go into the relationship with your eyes wide open. Don't take anything for granted, and check the individuals out for integrity and professionalism.

Where Should You Advertise?

Advertising in the local phone directory will generate customer inquires and provide credibility for your company. Ads in the local newspaper can result in quick responses. Door-to-door pamphlets and flyers can produce satisfactory results. Radio and television ads can be very effective, but they are expensive and require repetition. Putting a slide-in ad in the video boxes at the local rental store can give you a lot of exposure. The list of possible places to advertise is limited only by your imagination. However, some advertising media are better than others. Let's take a closeup look at some specific examples.

The Phone Directory

The phone directory is an excellent place to have your company advertised. The size of your ad, however, will depend on the nature of your business and the type of work you want to attract. Being listed in the phone book adds credibility to your company. Whether you are merely listed in a line listing or have a full-page display ad, you should get your company name in the phone book as soon as possible.

The size of your ad in the directory should be determined by the results you hope to achieve. Large display ads are expensive, and they may not pay for themselves in your line of work. If your business is building houses, a large display ad probably isn't necessary. When people are shopping for a builder, they are not normally in a hurry.

An ad that is one column wide and an inch or two in length can yield just as many calls. A quick look at how your competition advertises can give you a hint as to what you should do. If all the other builders have large ads, you probably should have a large ad.

Over the years I have tried many experiments with directory advertising. At one time I was running a half-page ad for my business. I thought I could save money by going to a smaller ad. I did pay less for my new ad, but my business suffered from the lack of the large display. I had a noticeable drop in phone requests.

As I became more knowledgeable about business, marketing, and advertising, I continued to test the results of various directory ads. During my test marketing, I used many types of ads for my various businesses. I found that for remodeling, real estate, and plumbing, large ads worked best. When perfecting my ads for homebuilding, I did just as well with smaller ads. The results of ad sizes have also varied geographically for me. My requirements in Virginia called for a bigger ad than in Maine.

PRO POINTER

The phone directory is an excellent place to have your company advertised. The size of your ad, however, will depend on the nature of your business and the type of work you want to attract. Being listed in the phone book adds credibility to your company. Whether you are merely listed in a line listing or have a full-page display ad, you should get your company name in the phone book as soon as possible.

Newspaper Ads

Newspaper ads provide quick results; you either get calls or you don't. As a service contractor, my experience has shown that most respondents to newspaper ads are looking for a bargain. If you want to command high prices, I don't think newspapers are the place to advertise. But if you are new in business, the newspaper can produce customers for you quickly.

Handouts, Flyers, and Pamphlets

Handouts, flyers, and pamphlets are similar to newspaper advertising. These methods seem to generate calls quickly, but the callers are usually looking for a low price on your services. Many businesses consider this form of advertising as degrading. I don't know that I would agree with that opinion, but I don't think you will receive the money you are worth with these low-cost advertising methods.

Radio Advertising

Radio advertising is expensive, but it is a good way to get your name to listeners. I believe the key to radio advertising is repetition. If you can't afford to sustain a regular ad on the radio, I would advise against using this form of advertising. Most people are not going to hear your ad and run to the nearest phone to call you. However, if you can budget enough money for several radio spots for a few weeks, you will gain name recognition.

Television Commercials

Television commercials can be very effective. People associate television advertisers with success. With the many cable channels available, television advertising can be an affordable and effective way to get your message out to the community. I have used ads on cable television very effectively. Television ads can increase your sales. Choose shows that are on home-improvement or do-it-yourself topics.

Direct-Mail Advertising

Direct-mail advertising can be very effective, but it is not always cost effective; because of the profit potential from building a new house, direct mail is usually a sound business decision in our field. The cost for direct-mail advertising can easily run into thousands of dollars. Most people who use this form of advertising are content if only one percent of the people they mail to become customers.

By using direct mail you can reach a targeted market. If you want to advertise to people with incomes in excess of $50,000, you can purchase a mailing

PRO POINTER

Direct-mail advertising can be very effective, but it is not always cost effective; because of the profit potential from building a new house, direct mail is usually a sound business decision in our field. The cost for direct-mail advertising can easily run into thousands of dollars. Most people who use this form of advertising are content if only one percent of the people they mail to become customers.

list of just those people. This type of demographic breakdown is very effective in mailing to the best prospects.

Most mailing lists are available at prices of around $75 for each thousand names. Many sellers of mailing lists require a minimum order of 3000 names. The names can be supplied to you on stick-on labels. Expect extra charges for various demographic breakdowns.

If you want to reduce your mailing costs, your local postmaster can provide a bulk-rate permit. To use the bulk-rate service, you must mail a minimum of 200 pieces of mail at a time. The cost for this type of mailing is much less than first-class postage, but there are one-time and annual up-front fees to be paid. You can consult your local post office for full details.

Creative Advertising Methods

Creative advertising methods are just that—creative. You might want to rent space on a billboard to advertise your business. Perhaps you will arrange a deal with a local restaurant to have your company highlighted on their menus. Providing uniforms for the local Little League can get your name in front of a large audience. If you put your mind to it, there is almost no end to the possibilities for creative advertising.

What Rate of Return Will You Receive on Advertising Costs?

The response to your advertising will depend on your marketing plan and the execution of your advertising. If you are advertising in the local newspaper, you might expect about a .001-percent response. In other words, if the paper has 25,000 subscribers, you might get 25 responses to your ad. This projection is aggressive, and in most cases your response will be much lower. If you advertise at the right time of the year with the right ad, 25 calls could come in. However, if you only get 10 calls, don't be surprised. In some cases you may not even get 10 calls. Your adver-

PRO POINTER

The response to your advertising will depend on your marketing plan and the execution of your advertising. If you are advertising in the local newspaper, you might expect about a .001-percent response. In other words, if the paper has 25,000 subscribers, you might get 25 responses to your ad. This projection is aggressive, and in most cases your response will be much lower.

tising success will depend entirely on how well you picked the publication and designed the ad.

Advertising a contracting business on the radio or television can seem like a waste of money. It is not uncommon for these ads to run without getting calls, but that doesn't mean that the ads were not effective. Television and radio advertising builds name recognition for your company. This form of advertising works best when it is used in conjunction with some type of print advertising.

If you are running ads in the paper, distributing flyers, or doing a direct-mail campaign while the television and radio ads are on, you should see a higher response than you would without the radio and television ads.

PRO POINTER

Direct-mail advertising often provides fast results. Many people receiving ads by mail either trash them or act on them quickly. A one-percent response on direct-mail advertising is generally considered good. For example, if you mail to 1,000 houses, you should be happy if you get 10 responses. Due to the low response rate of bulk mailings, direct-mail is not effective for low-priced services. However, if you are selling big-ticket items such as new houses, direct mail can work very well.

Direct-mail advertising often provides fast results. Many people receiving ads by mail either trash them or act on them quickly. A one-percent response on direct-mail advertising is generally considered good. For example, if you mail to 1,000 houses, you should be happy if you get 10 responses. Due to the low response rate of bulk mailings, direct-mail is not effective for low-priced services. However, if you are selling big-ticket items such as new houses, direct mail can work very well.

If you target your direct-mail market, you should do much better on your rate of return. For example, if you use a mixed mailing list for your advertising, you don't know what type of person is receiving your ad. But if you pick a list based on demographics, you can be sure you are reaching the type of potential customer you want.

Demographics are statistics that tell you facts about the names on your mailing list. You can rent a mailing list that consists of specific age groups, incomes, and so forth. These statistics can make a big difference in the effectiveness of your advertising.

Determining the effectiveness of advertising is a task all serious business owners must undertake. To learn which ads are paying for themselves, you need to know which ads are generating buying customers. Some ads generate a high volume of inquiries but don't result in many sales. Other ads produce fewer curiosity calls and more buying customers. You need to track the results of your advertising. Without knowing which ads and advertising media are working, you have no way of maximizing the return on your advertising expenses.

Learn to Use Advertising for Multiple Purposes

Most businesses learn to use advertising for multiple purposes. The primary use of advertising is to generate consumer interest in goods and services. But advertising can do much more for a business. As we have already seen, advertising can build name recognition for your company. Name recognition is important when trying to get the most mileage out of your advertising budget.

Advertising can be used to build your company image. Through advertising, you can create almost any look you like for your business. A company image can be responsible for commanding higher fees and quality customers.

Advertising can enable you to establish your goals. If you want to be known as an expert in restoring old homes or building authentic reproductions, advertising can get the job done. As you go along in business, you will find that various forms of advertising can help you achieve success in many ways.

Building Name Recognition through Advertising

We have already talked about building name recognition through advertising, but now we are going to learn how it's done. You want people to see or hear your company name and feel like they know the company. To accomplish this goal, you must use repetitive advertising.

Repetitive advertising can be used in all formats. Take radio advertising, for example; when you hear radio commercials, you normally hear the company name more than once. Pay attention the next time you hear ads on the radio. You will probably hear the company name or the name of the product being sold at least three times.

Television uses verbal and visual repetition to plant a name or product in your mind. Watch a few television commercials and you will see what I mean. During the commercials you will see or hear the company name or product several times.

PRO POINTER

You want people to see or hear your company name and feel like they know the company. To accomplish this goal, you must use repetitive advertising.

Not only should your name be used often in the ad; the ad should also be run regularly. If you advertise in the newspaper, don't run one ad and stop. Run the same ad several different times. Use your logo in the ad, and keep the ads coming on a regular schedule. This type of repetition will implant your company name into the subconscious of potential customers. When these potential customers are ready to become actual customers, they will think of your company.

PRO POINTER

If you advertise in the newspaper, don't run one ad and stop. Run the same ad several different times. Use your logo in the ad, and keep the ads coming on a regular schedule. This type of repetition will implant your company name into the subconscious of potential customers.

Generating Direct Sales Activity with Advertising

The need for generating direct sales activity with advertising is the reason most people use it. For a service business, generating direct sales is possible with direct mail (print or email), radio, television, print ads, telemarketing, and other forms of creative marketing. Telemarketing and direct mail are two of the fastest ways to generate sales activity.

We've already talked about how direct mail works, but how about telemarketing? Telemarketing is a tough job. Calling people you don't know and asking them to use your services, buy your product, or allow you into their home for a free inspection, estimate, or whatever is not much fun. However, if you can live with rejection and are not afraid to call 100 people to get 10 sales appointments, cold calling will work.

When your objective is to generate sales activity, it helps to make your offer on a time-restricted basis. By this I mean that you should offer a discount for a limited time only. Create a situation where people must act now to benefit from your advertising. Time-sensitive ads can generate activity quickly.

Without Advertising, the Public Will Not Know that You Exist

Without advertising, the public will not know that you exist. Advertising is expensive, but it is also a necessary part of doing business. If you don't spend money on advertising, the public is not going to spend money with your business. The contracting field is filled with business owners who are aggressive. These aggressive owners advertise regularly. If you don't put your name in front of people, you will be run over by the companies that do.

Promotional Activities

Promotional activities are an excellent way to generate more sales and to build name recognition. By using special promotions you capture public attention and create an opportunity for additional sales. Let me give you an example of how you could stage a promotional event.

PRO POINTER

Promotional activities are an excellent way to generate more sales and to build name recognition. By using special promotions you capture public attention and create an opportunity for additional sales.

In this example, assume that you are a contractor who specializes in new homes. You could talk to your local materials supplier and develop a seminar. Ask the supplier to allow you to come into the store and give a homebuilding or homebuying seminar for shoppers. Tell the supplier how the seminar will be good for the store's image and increase sales.

Advertise the free seminar for about two weeks prior to the date of your talk. The supplier may be willing to pay a portion of the ad costs—after all, the store is gaining publicity from this promotion as well.

When people begin gathering around you in the store, be sure to have your business cards, rate sheets, and other sales aids displayed where the shoppers can see them. After your seminar, field questions from the audience. This type of promotion can create the image of you as an expert in your field.

If it is legal in your area, give away a door prize. Have the audience fill out cards with their names, addresses, emails, and phone numbers for a prize drawing. Give away the prize; it could be a discount on remodeling services, a small appliance, or just about anything else you can think of. After the seminar, you have a box full of names and addresses to follow up on for work. This type of idea can increase your business dramatically.

How to Stay Busy in Slow Times

Every business owner wants to know how to stay busy in slow times. Until you have survived a business recession, you may not have the experience to stay afloat in troubled waters. Since you can't always learn survival skills on a first-hand basis and survive, you must turn to the experience of others for your training.

When times are tough, you may have to alter your business procedures. How will you do it? Will you lower your prices? Will you eliminate some overhead expenses? How about cutting back on your advertising expenses? Any of these options could be the wrong thing to do.

If you lower your prices, you will have a very hard time working your prices back up to where they were. Lowering prices is risky business, but sometimes it is the only way to keep food on the table. If you have to lower prices to stay in business, do so with the understanding that returning prices to normal will take time.

Discounts might accomplish the same goal as lowering basic prices but with fewer long-term effects. People expect discount offers to end. Run ads offering a discount from your regular labor rates for a limited time only. This tactic will be less difficult to deal with later than a full-scale lowering of labor rates.

Cutting unnecessary overhead expenses is a good idea at any time, but cutting needed overhead expenses can be dangerous. Even when times are rough, some overhead expenses should be maintained. If you cut back on the wrong expenses, your business may get worse. For example, if you eliminate your answering service, you may miss some of the important business calls your company needs to survive. Evaluate your expenses carefully before making cuts. Only eliminate expenses that will not affect the volume and profitability of your business.

Advertising is one of the first expenditures many business owners cut. Advertising is just as vital to your business in bad times as it is in good times. In fact, advertising may be more valuable in tough times. As your competitors cut back and shrink into obscurity, you can lunge forward with aggressive advertising. Think long and hard before you eliminate your ad in the phone book or reduce your normal advertising practices.

Business promotion should be one of your top priorities. As I've said before, without customers you have no

PRO POINTER

Advertising is one of the first expenditures many business owners cut. Advertising is just as vital to your business in bad times as it is in good times. In fact, advertising may be more valuable in tough times. As your competitors cut back and shrink into obscurity, you can lunge forward with aggressive advertising. Think long and hard before you eliminate your ad in the phone book or reduce your normal advertising practices.

business. To get customers, you must always strive to find new people to build homes for. If you get too comfortable and slack off on your business promotion, you may find that your business will shrivel up and blow away.

Green Land Developing Could Double Your Income

Creating green subdivisions could be your ticket to higher income. It is not unusual for builders to become their own land developers in order to double their money on building lots. How is this possible? The builder who develops lots gets the development profit and the building profit. Old-school developers try to double their money on lots. For example, if a lot costs $50,000 to develop, the sales price to the public would ideally be around $100,000.

When builders buy finished building lots, they pay a premium price. The cost might be a little less than what the general public would pay, but the land cost has a lot of profit built into it. Buying the lot instead of developing the lot removes your ability to cash in on the extra profit. There are heavy risks involved in land development, so don't get the wrong impression. This is not a suitable business venture for inexperienced builders who are not willing to invest substantial time in a learning curve.

Developing your own building lots can sound like a good way to control your own subdivision and make a lot of extra money at the same time. The truth is that developing land can result in these types of advantages as well as others. On the flip side, getting into land development can be very risky. Unless you've worked with a developer, a surveyor, or someone else who has shown you the ropes, entering into land development can be very dangerous financially.

When I first entered the trades, I worked for various developers who were also builders. The first two people of this type whom I worked with handled large developments. My first experience was with single-family homes and lots. When I changed jobs, I was introduced to a 365-unit townhouse development. Even though my work didn't include development duties, I got to see a lot of what goes on.

When I went into business for myself, I never contemplated becoming a developer. My goal was plumbing and remodeling. Even when I started building houses, I never gave much thought to being a developer. Then I ran into a developer who was aging and looking to sell out his inventory of land. Some of his property was already developed into ready-to-build lots, and some of it was still in its raw stage. This was when I first thought seriously about trying to develop my own lots.

PRO POINTER

Developing your own building lots can sound like a good way to control your own subdivision and make a lot of extra money at the same time. The truth is that developing land can result in these types of advantages as well as others. On the flip side, getting into land development can be very risky. Unless you've worked with a developer, a surveyor, or someone else who has shown you the ropes, entering into land development can be very dangerous financially.

At the time my first development opportunity came along, I was young and ambitious. But I was also nervous about testing the waters of land developing. After hearing horror stories and talking to developers who didn't succeed, I decided not to pursue the land deal. It might have been a mistake, but I believe I did the right thing.

About a year after I declined my first development opportunity, I ran into a developer who used the same survey and engineering firm that I did. We talked on occasions and got to know each other one week at a time. I built a new home for one of the partners in the surveying firm, and that got me closer to the inner circle. Before long, I was talking about development deals with the developer I had met. And fairly soon we were working together. The development projects included commercial space for a shopping center, in-town land for housing, and rural land, where we created 10-acre mini-farms.

I've been fortunate with my development deals. So far I've not delved in too deep or lost serious money on land deals. Based on what I know of the industry, I'm an exception rather than the rule. Buying and developing raw land is risky. But a person's reward is often in direct relationship to the amount of risk taken.

Land development can be as simple as buying a tract of land and cutting it in half. Complex development deals can take years to see results. The approval process that generally is required for major developments can be very lengthy and expensive. If you don't have deep pockets, big development deals are not a good venture to become involved with.

Little Deals

Little deals are a good place to start as a developer. Buying some land and cutting it into a few building lots is something that most builders can handle if they put their minds to it. The cost for this type of developing is usually low in comparative terms. Let's talk about some of the potential expenses that you will have to be prepared to handle if you want to create your own building lots.

PRO POINTER

Little deals are a good place to start as a developer. Buying some land and cutting it into a few building lots is something that most builders can handle if they put their minds to it. The cost for this type of developing is usually low in comparative terms. Let's talk about some of the potential expenses that you will have to be prepared to handle if you want to create your own building lots.

Survey and Engineering Studies

Survey and engineering studies are usually among the first expenses incurred once a developer has found a piece of land that is intriguing. Surveys are done for obvious reasons. You want to know how much land you're buying and what shape it is in. If the land is in a flood-zone or plane, a thorough survey will reveal it. Elevations will be included on comprehensive surveys. This shows the grade of roads that must be built, the drop-off of house foundations, and so forth. A full-blown survey will probably cost anywhere from several hundred dollars to a few thousand dollars. The cost depends on past surveys, land size, geographic location, and other factors.

The engineering studies can be extremely expensive, or they can amount to only a few thousand dollars. It depends upon what you are looking for. If you just want to establish approved perk sites for septic systems, the cost will not be staggering. When you want to know soil consistencies, compaction rates, water-retention details, and so forth, the cost will likely reach well up into many thousands of dollars.

Engineering firms often represent developers during the approval process of large projects. Paying professional hourly rates for meeting after meeting, month after month, can run into some major money. This won't normally be needed on little deals, but it is another potential expense to be aware of.

Acquisition

Once you have had the property checked out from north to south, you will need money or credit to secure the land. Down payments on land are often required to be at least 20 percent of the land's market value, and a 30-percent down payment is common. So, depending on the price of the land, this can amount to quite a bit of cash. But there is often a way around this problem.

Sellers of land will frequently finance a sale themselves. If you're getting owner financing, the down payment could be 10 percent or less. The last piece of land I bought was conveyed to me with a few hundred dollars in the form of a deposit and interest-only monthly payments. When you are cutting a deal with a seller who has clear title to the land, you can make some very creative deals.

PRO POINTER

Once you have had property checked out from north to south, you will need money or credit to secure the land. Down payments on land are often required to be at least 20 percent of the land's market value, and a 30-percent down payment is common.

Filing and Legal Fees

Filing fees, legal fees, and other costs associated with a development deal can run up to a sizable sum. The largest portion of these expenses is normally the legal fees, closing costs, and points. If your deal is a simple one and you are financing the property with the owner, your costs could wind up being less.

Hard Costs

The hard costs of land development vary with the type of land being developed and its location. If you have to pay to have electrical power, water mains, or sewers extended to the land, you could be looking at thousands of dollars. If utilities are already available, the cost is much less. In the case of rural property where wells and septic systems will be used, you don't have to worry about water and sewer fees. The builder will have to foot the bill for a well and septic system.

Clearing a piece of land is usually left for a builder to do. Developers sometimes have to pay for some clearing to allow the installation of roads, but it is unusual for a developer to prepare lots to the point where house sites are cleared.

Installing a road, even a gravel road, can be very expensive. The cost of a road can be enough to make a development deal go sour. This expense, like others, should be determined while you still have an escape clause in your contract. The last thing you want is to buy a piece of land fast and then find out that the cost of developing it makes the venture too expensive to proceed with. You will be left with a piece of raw land that is worth no more than what you paid for it, and you'll have to bear the burden of selling it. Get estimates on all your projected development costs prior to making an ironclad commitment to purchase land. Your engineering firm can help you to identify the types of costs to consider.

PRO POINTER

Installing a road, even a gravel road, can be very expensive. The cost of a road can be enough to make a development deal go sour. This expense, like others, should be determined while you still have an escape clause in your contract. The last thing you want is to buy a piece of land fast and then find out that the cost of developing it makes the venture too expensive to proceed with.

When Things Go Right

When things go right, developing raw land into a small number of house lots is not too complicated. Let me give you an example of how simple it can be. I bought 18 acres of rural land. If I wanted to cut this acreage into two building lots, say, one 8-acre parcel and one 10-acre piece, I could do it for less than $1,000. The land would have to be perk-tested to be sold as building lots, but this would not be very expensive. In my case, a formal survey would not be needed at the time of development. Using town tax maps, I could draw my own plot plan. Legal expenses would have to be considered. The land has road frontage and available electrical service, so there's no cost in these categories. Filing fees with the town would be minimal

If I chose to divide my 18 acres into more than two lots, the game would change. Creating more that two lots falls under different rules and regulations. My engineering costs would escalate, as would most of my other expenses. The amount of time needed to gain approval for a larger subdivision of the land could exceed one year. In theory, I should make more money by having an additional lot to sell, but I might not. Between the time lost, the extra money spent, and the fact that the lots would be smaller, I might not make as much money, and I would certainly have to endure a great deal more frustration. By keeping your first few deals simple, you can get a taste of what developing is like, and if you're fortunate, you can make enough money to fund larger deals when you are prepared to do them.

Mid-Size Deals

Mid-size deals I usually consider to be a no-man's-land. This is one reason why I passed on my first development opportunity. When you do a little deal, your costs are low and your profits tend to be

PRO POINTER

Mid-size deals I usually consider to be a no-man's-land. This is one reason why I passed on my first development opportunity. When you do a little deal, your costs are low and your profits tend to be high in terms of percentages. Big deals that work produce fabulous profits, and there is enough profit potential in a big deal to warrant an all-out promotional campaign. Mid-size deals, such as a 12-lot subdivision, require a lot of money to get going, and they don't produce enough profit to justify major ad campaigns and publicity.

high in terms of percentages. Big deals that work produce fabulous profits, and there is enough profit potential in a big deal to warrant an all-out promotional campaign. Mid-size deals, such as a 12-lot subdivision, require a lot of money to get going, and they don't produce enough profit to justify major ad campaigns and publicity. This is why I see them as a type of quicksand for developers. You can get sucked into a bad situation very fast when you take on a mid-size deal.

I can't tell you that you can't or won't enjoy success with medium-sized developments. Some developers do okay with them, but I think the odds are against you. There are a few expectations to my rule. One of the deals I was involved in consisted of about 200 acres of land. We cut the parcel into twenty 10-acre lots, built a spec house on the most visible lot, and promoted the properties as mini-farms. The land was located within a 20-minute drive of a major city, and the deal worked well. I would consider this a mid-size deal, and it paid off. But we had a lot going for us. The location was good, the cost of development was affordable, with the road being our biggest expense, and our package price for house and land was low. This combination led us to victory. Still, I'd be very careful about doing a deal like this one.

Big Deals

Big deals can make you dream of early retirement. The amount of money that can be made by developing a shopping center or a large housing development is staggering. Unfortunately, so is the cost of preparing the project for sale. I don't have a lot of first-hand experience as a major developer, so I won't attempt to tell you all the ins and outs. I simply don't know them. As a part of a team, I have worked with big development deals. I saw enough to know that I did not have enough money or experience to tackle such a task alone.

Millions of dollars can be made on big development deals, but a developer needs a tremendous amount of money or backing to pull off the process. And the deal could go bad, causing everyone involved to wind up bankrupt. I'm biased by a lack of knowledge, but I would advise you to avoid big deals unless you can do them like I did and work with someone who knows what to do and when to do it.

It's Tempting

It's tempting when you see a parcel of land for sale that you feel would make a good subdivision. You can draw a plot plan and count the number of lots you would get out of the property. Then you will probably punch numbers into your spreadsheet program to see how much money you would make by doing the deal. This is fun and the numbers can make your pulse rate climb, but you have to be realistic.

Installing roads eats up buildable land. Some builders don't take this into consideration when mapping out a development plan. Retention ponds, common space, and other zoning requirements can diminish the amount of land left to build on. What looks like a 25-lot subdivision might yield only 15 lots. There's a huge difference in the profit potential between 15 lots and 25 lots. Take your time and make sure your

PRO POINTER

It's tempting when you see a parcel of land for sale that you feel would make a good subdivision. You can draw a plot plan and count the number of lots you would get out of the property. Then you will probably punch numbers into your spreadsheet program to see how much money you would make by doing the deal. This is fun and the numbers can make your pulse rate climb, but you have to be realistic.

calculations and estimates are correct. It's very easy to make mistakes in land development that can put you out of business entirely.

Going Green

Going green is the wave of the future. If you venture into land development, study your options closely. Keep the development as environmentally friendly as you reasonably can. Look into retention ponds that could be used to provide recycled water for irrigation purposes. Build in suitable green space for nature trails, natural flora and fauna, and so forth. You might want to pave every available acre, but this could be a major mistake when compared to the potential value you could have with a green development.

Take your time. Go slow. Learn all that you can before you jump into land development. Once you feel that you are ready, test the waters a little at a time. Don't get greedy. If you can establish yourself as a green developer, you will have a much higher profit potential.

PRO POINTER

Going green is the wave of the future. If you venture into land development, study your options closely. Keep the development as environmentally friendly as you reasonably can. Look into retention ponds that could be used to provide recycled water for irrigation purposes. Build in suitable green space for nature trails, natural flora and fauna, and so forth.

Green Landscaping Tips That Sell Houses Fast

Experienced builders know that landscaping is a key element in selling houses. Curb appeal counts. It sounds a bit strange to talk about green landscaping. Is there really a way to landscape with special techniques to keep the work green? Yes. Sustainable landscaping is well known in the industry, and it plays several roles in building.

What is green landscaping? There are a number of ways to describe it. Organic materials are one aspect. Erosion control is one of the benefits of green landscaping. Did you know that green practices can include the use of a vegetated roof? I'm not kidding. Grow your own roof and benefit from lower heating and cooling expenses. Water conservation can come from green landscaping. Yes, there are a number of ways to incorporate special landscaping in green buildings.

Conservation

The conservation of natural resources is a very good reason to use green landscaping practices. Our resources are limited. Building projects that enable property owners to minimize the depletion of water, soil, and fossil fuels is a worthy goal. Customers appreciate builders who provide opportunities to live responsibly in ways that do the least harm to the environment.

Water

Water conservation is important to all of us. We need clean water to survive. How much water is used to irrigate plants and water lawns? A lot, that's how much. Government statistics indicate that water usage for irrigation in the East accounts for approximately 30 percent of all water used. In the West the percentage jumps to about 60 percent. Think about this. Somewhere near one-half of all the water consumed in the United States is used to water lawns and plants.

When droughts strike, water becomes very precious. This has been experienced time and time again. Not long ago, entire states in the South were in danger of running out of potable water. What can you and caring customers do about this? Most of us can't make it rain, but we can landscape and build in ways to reduce the amount of clean water required to keep vegetation healthy.

You have to choose plants that are suitable for the region in which you are building. A few factors to consider include the following:

- Climate
- Soil types
- Sunlight
- Exposure
- Topography

When plants are placed close together, you can conserve more water. The proximity of the plants to each other makes the watering area smaller. Summer months are the worst time to plant. Depending on the types of planting being done, spring and fall make for better plant health.

Use recycled water or storm water that has been stored to keep landscaping plants healthy. Less water will be wasted when plants are watered in the early morning

PRO POINTER

Water conservation is important to all of us. We need clean water to survive. How much water is used to irrigate plants and water lawns? A lot, that's how much. Government statistics indicate that water usage for irrigation in the East accounts for approximately 30 percent of all water used. In the West the percentage jumps to about 60 percent. Think about this. Somewhere near one-half of all the water consumed in the United States is used to water lawns and plants.

hours. When it is cooler and less sunlight is available, water will not evaporate as quickly. A drip irrigation system can be used to reduce water waste. Rain gauges and timers are effective tools in water conservation.

Soil

How often have you considered the soil under your boots? It is common for people to assume that dirt runs deep. This is not really the case. Soil is actually limited. Should soil be considered a renewable resource? Yes. The problem is that it takes a very long time for new soil to develop. Years often pass as rocks break down and organic matter decomposes. This is necessary for the creation of natural soil.

How deep does dirt go? In worst cases, soil can be less than one inch deep. In well developed areas, such as farms, the depth can run several feet. The wide range of depths requires developers to test soil levels.

Erosion is a major concern with soil cover. Preventing erosion translates into conserving the earth, and I mean this very literally. Slopes, banks, and steep areas are subject to substantial erosion risk. Planting in these areas can save the soil. Building strength with turf will help to stabilize soil. But the shallow roots of grass are not always enough. Choose plants with deeper roots. Groundcovers and other plants should be installed close together. Mulch the landscaped area and keep it mulched until the plants fill in the open areas. Use trees and shrubs where you can. The roots of these plants run deeper and hold the ground in place better than shallow-root plants.

PRO POINTER

How deep does dirt go? In worst cases, soil can be less than one inch deep. In well developed areas, such as farms, the depth can run several feet. The wide range of depths requires developers to test soil levels.

Fossil Fuels

Fossil fuels include coal, gas, and oil. As you no doubt know, these fuels are finite in their supply. The conservation of fossil fuels is more of a responsibility for property owners than it is for green builders. For example, homeowners can use hand tools and composting to reduce energy usage. As a builder, you can do your part by integrating natural landscape design in your projects.

Natural Designs

Natural designs in landscaping are becoming more and more popular. There are good reasons for this. How does natural design compare with traditional landscaping? Keeping the explanation simple, it can be said that a natural design is casual, and traditional landscaping is formal. Due to this difference, old-school landscaping requires more time, attention, and money to maintain. Consider the following pros and cons for each type of landscaping:

- Traditional landscaping calls for careful pruning.
- Traditional landscaping is usually installed in static rows or patterns.
- Natural landscaping blends in with the surroundings.
- Natural landscaping incorporates the use of native plants rather than expensive exotic offerings.
- Man-made garden decorations are often present in traditional landscaping.
- Natural landscaping utilizes rocks, trees, logs, and even shallow ponds.
- The random design of a natural garden reduces the amount of imperfections that would show up in a formal garden. This reduces maintenance.
- Fertilizer and pesticides are frequently used to maintain the formal look of traditional landscaping. This is not needed nearly as often with a natural garden.
- Natural designs tend to reduce the requirements of maintenance, the use of chemicals, and the need to use power equipment to keep up the good look of the garden.

Working with Native Plants

Working with native plants in your landscaping offers several benefits. Naturally, the cost of these plants is usually less than the expense of exotic plants. The success rate for good growth is better for native plants. It is surprising how few native plants are used in major landscaping projects.

In general, native plants typically require less fertilizer, less maintenance,

PRO POINTER

In general, native plants typically require less fertilizer, less maintenance, and fewer pesticides. Any building uses up natural space. This affects birds, animals, insects, and other natural creatures. It is only fair to give these affected species a natural landscape to come to as an oasis in a sea of pavement.

and fewer pesticides. Any building uses up natural space. This affects birds, animals, insects, and other natural creatures. It is only fair to give these affected species a natural landscape to come to as an oasis in a sea of pavement.

As a builder, you can landscape in a way to attract birds, butterflies, and beneficial insects. Many customers will see this as an advantage over the house down the street that offers skimpy landscaping. It is not only possible to enhance a building project to be more harmonious with nature; it is a requirement for good green builders.

Cutting Heating and Cooling Costs

You can use creative landscaping while cutting heating and cooling costs for your customers. Planning a solid planting strategy in advance can benefit property owners for many years to come. A study done in Pennsylvania found that a house shaded by trees could reduce the cost of air conditioning by as much as 75 percent. That is considerable.

Common practice calls for deciduous trees, the ones that lose their leaves in winter, on the west and south sides of buildings. These directions receive the most intense sunlight and heat. The leafy trees provide much needed shade in warm months. When the leaves fall to the ground to be composted in the fall, the bare branches allow the warming sun to help heat a building. Reports have indicated that communities that contain shade trees benefit from cooler summer temperatures--as much as six degrees cooler.

Winter winds tend to come from the north and the west. Consider planting evergreen trees and shrubs in locations that will protect your buildings from these winds in the winter. Heating costs should go down. Evergreen trees are normally planted well away from a building. Determine the mature height of the trees you are planting. When building a windbreak, multiply the height by between two to five times--that is how far from a foundation the trees should be placed.

Evergreen shrubs can be planted around a foundation to block wind exposure. Be sure to install the shrubs far enough away from the building to avoid moisture

PRO POINTER

You can use creative landscaping while cutting heating and cooling costs for your customers. Planning a solid planting strategy in advance can benefit property owners for many years to come. A study done in Pennsylvania found that a house shaded by trees could reduce the cost of air conditioning by as much as 75 percent. That is considerable.

problems in the future. Exact planting distances will be determined by the type of shrub being used.

What do you think about planting a roof on a home? It happens. Green roofing requires special construction design, but the procedure provides strong insulation value to reduce utility costs. I am told that these roofs enjoy a long life. I am not sure I would want to climb up on the roof to cut the grass, but it is one more way to go green.

Flooding

Paving developments can lead to flooding. With flooding you can experience erosion, stream pollution, destruction of aquatic habitat, and property damage. Builders can work with developers to reduce the impacts that contribute to flooding. Green developers will already be aware of this, but as a builder you may have to seek out green developments to house your buildings.

PRO POINTER

Paving developments can lead to flooding. With flooding you can experience erosion, stream pollution, destruction of aquatic habitat, and property damage. Builders can work with developers to reduce the impacts that contribute to flooding.

Here are a few tips on how to build with the risk of flooding in mind:

- Avoid using concrete, mortared brick, and stone patios. Instead, build a wooden deck or use patio blocks that are not mortared together.
- Paved driveways are often desirable, but they cause water runoff. The same is true for concrete driveways. Installing a gravel driveway allows water to seep into the ground below the gravel. A pervious asphalt drive is another option. Parking areas account for substantial water runoff.
- Avoid paved walking paths. Use gravel or mulched paths as a replacement.
- Lawns are good, right? Yes, but native groundcovers, planting beds, and trees and shrubs can do a lot more to control erosion and water runoff.

Storm Water

Storm water can be an enemy. Equip your homes with a good gutter system. Where allowed by law and code, fit the downspouts with buried drain pipe that will divert rain water to either a French drain, a grassy swale, or a rain garden.

What is a rain garden? It is a small retention pond. You can make one by creating a depression in the ground to accept runoff water. Plant the area with shrubs and other plants that are able to tolerate wet conditions. The rain garden will collect the water and allow it to perk into the earth slowly. Before doing this, make sure that the soil will perk satisfactorily to avoid creating a full-time pond.

PRO POINTER

Storm water can be an enemy. Equip your homes with a good gutter system. Where allowed by law and code, fit the downspouts with buried drain pipe that will divert rain water to either a French drain, a grassy swale, or a rain garden.

Rain barrels have come back into fashion. These barrels can collect water from a gutter system and retain it for watering landscaping. Commercial rain barrels are equipped with valves and hose threads.

Sump pumps, when required, should be drained to a suitable area, such as a French drain or rain garden, subject to local code requirements. Storm-water terminals should be kept at least 10 feet from foundations to avoid moisture problems with buildings.

Selling Your Advantages

As is common in most competitive marketplaces, you should consider all the options you have for selling your advantages. The fact that you are a green builder is a good start. What sets you or your projects apart from your competitors? Find these advantages and a way to communicate them to customers, and you should have a winning combination.

Will you plan your own landscaping details? Should you bring in a landscape architect? Can advisors from the local nursery help you create a green landscaping plan? How you go about determining what to plant and where to plant it is up to you. On large projects, a landscape architect is a good resource to consider. Researching land-

PRO POINTER

As is common in most competitive marketplaces, you should consider all the options you have for selling your advantages. The fact that you are a green builder is a good start. What sets you or your projects apart from your competitors? Find these advantages and a way to communicate them to customers, and you should have a winning combination.

scaping techniques on the Internet can provide you with ideas, options, and answers. The vendors who will supply you with plants are generally an excellent source of information.

Invest the time and money needed to make a viable landscaping plan. What is the first impression that you form when driving up to a home that is for sale? Many factors are considered. You will probably take note of the house style, the type of roofing used, the color of siding, doors, and trim, and so forth. Without even realizing it, you are almost certain to be swayed by the landscaping. It may be subliminal for some people, but it counts and it counts big.

PRO POINTER

Invest the time and money needed to make a viable landscaping plan. What is the first impression that you form when driving up to a home that is for sale? Many factors are considered. You will probably take note of the house style, the type of roofing used, the color of siding, doors, and trim, and so forth. Without even realizing it, you are almost certain to be swayed by the landscaping. It may be subliminal for some people, but it counts and it counts big.

If you are selling spec houses, the exterior finish, including landscaping, is a key element in convincing potential buyers to come inside and look around. Model homes in subdivisions should be dressed up nicely with landscaping. However, avoid the bait-and-switch move of loading a model home with every available option and then upselling the customer from a base price to a much higher price to get what the customers thought they were looking at in the first place. People don't appreciate this type of sales approach.

There is no doubt in my mind that landscaping is one of the first steps in selling a finished home on the open market. I have built and sold a lot of homes. In the beginning I did minimal landscaping. With time and experience, I changed my practices and I enjoyed better sales. Do not overlook the value of landscaping.

FIGURE 15-1 Attractive lanscaping can make the difference in whether a home sells (courtesy of ECO-Block, LLC).

Glossary of Green Words and Terms

Absorption: The process by which light energy is converted to another form of energy, such as heat.

ACH (Air changes per hour): The movement of a volume of air in a given period of time. An ACH rate of 1.0 means that the volume of air will be replaced in a period of one hour.

Acid leachate: Water that is made highly acidic after seeping through landfills; it may be harmful to fish habitats and drinking-water supplies.

Active system: Heating, cooling, and ventilation systems that condition the air supply in buildings by using electricity or gas power.

Adaptable buildings: Buildings that may easily be remarketed, reconfigured, or retrofitted in order to meet the changing needs of maintenance crews, occupants, and the surrounding community.

Adsorption: The process in which the molecules of a gas, liquid, or dissolved substance adhere to a surface.

AFUE Annual fuel-utilization efficiency: A measurement of the efficiency of gas appliances, the ratio of energy output to energy input on an annual basis.

AFV (Alternative-fuel vehicle): Any vehicle powered by fuels other than gasoline.

Agricultural by-products: Materials left over from agricultural processes, such as shells and stalks, now being used for building materials.

Air infiltration: Undesired air leakage in a building shell. Air leaking out is referred to as exfiltration.

Air retarder/air barrier: Materials installed around the building frame to prevent and reduce air infiltration, used to increase energy efficiency by keeping out air that may be too cold, hot, or moist.

Albedo: The ratio of light falling on a surface to the amount of light reflected off the surface. Roofing materials with different ratios (high or low) of albedo produce different results.

Appraisal value: Generally substantiated through comparison with other properties, the estimated value of a property.

Allergen: Any substance that a person may show an allergic reaction to.

Attic venting system: Ventilation devices installed to allow fresh air in and allow the exhaust of air out to control air quality in an attic. The use of a continuous soffit vent with a continuous ridge vent is most effective, because it allows for even air flow along the underside of the roof while exhausting the hottest air at the highest point in the attic.

Autoclaved cellular concrete: A molded mix of concrete, sand, lime, water, and aluminum, which is then steam-cured for strength. Benefits of this process include a non-combustible, easily worked product.

Backdrafting: Combustion gases entering the living space instead of being properly drawn up an exhaust pipe; this happens as a result of depressurization, sometimes caused by exhaust fans or furnace heat.

Backflow preventer: In accordance with many building codes, an anti-siphoning device used on water pipes to prevent contaminated water from backing up into the water system.

Balance point: The temperature at which a building's internal heat gains are equal to the building's loss of heat to the surrounding environment.

Ballast: A magnetic or electronic device used to provide the stabilizing current or starting voltage of a circuit.

Bioaccumulants: Found in contaminated air, water and food, substances that are very slowly metabolized or excreted by a living organism, causing buildup over time.

Biodiversity: An ecosystem that contains a great variety of species and a complex web of interactions between them. Human influence, mainly related to resource consumption, greatly reduces biodiversity and heightens risk of catastrophic disruption to these systems.

Bioengineering: Using a combination of living plants and nonliving materials to stabilize slopes and drainage paths.

Biological wastewater management: Wastewater purification programs that are based on natural processes of wetland environments and powered by sunlight and microscopic living organisms.

Biomass: Any natural material—animal manure, wood and bark residues, or plants and plant materials.

Biomass energy: Energy released from biomass as it is used or converted into fuel.

Blackwater: Dirty water from sources such as toilets, kitchen sinks, and washing machines, which may be contaminated with harmful bacteria or microorganisms.

Blown-in batt: A method that uses a high-power blower and a fabric containment screen to install loose insulation into wall cavities.

Borate-treated wood: A mineral derived from borax that is less harmful than most other wood treatments. Treating wood with borate makes it more resistant to moisture and termites.

Brownfields: Areas where environmental contamination inhibits the redevelopment or expansion of old commercial and industrial sites.

BTU (British thermal unit): A measurement of heat energy, about the amount of heat needed to raise the temperature of a pound of water by one degree, or the amount of energy released by lighting a match.

Building codes: Municipal ordinances pertaining to health and safety, used to regulate construction and occupancy rules.

Building ecology: The physical environment of the interior of a building. Air quality, acoustics, and electromagnetic fields are key issues of building ecology.

Building envelope: The enclosed space of a building, closed in by walls, windows, roofs, and floors, through which energy is transferred to and from the interior and exterior.

Built environment: Any human-built structure in contrast to the natural environment.

Caliche: A common roadbed material made of calcium-carbonate-rich soil, which hardens without firing.

Capitalization rate: The rate at which future income flow is converted into a present value figure, expressed in a percentage.

Carbon dioxide (CO_2): Composed of one carbon atom and two oxygen atoms, a molecule formed through the processes of animal respiration or in the decay or combustion of organic matter. It is an atmospheric greenhouse gas, and plants absorb it in the process of photosynthesis.

Carbon monoxide (CO): Formed as a product of the incomplete combustion of carbon, a very toxic gas made of carbon and oxygen. When burned, it shows a blue flame and creates carbon dioxide.

Carrying capacity: The amount of a particular product that may be used without depleting the source or degrading dependent life forms.

Cellulose: The fibrous part of a plant, currently used to make paper and textiles, and may also be used to make building products such as insulation.

Cementitous: Having the properties of cement, the primary bonding agent in concrete.

Certified lumber: Lumber that is harvested from operations certified as sustainable.

CFCs (chlorofluorocarbons): A number of compounds used in refrigeration, cleaning solvents, and aerosol propellants that contain carbon, chlorine, fluorine, and hydrogen. CFCs are also used in the manufacturing of plastic foams and have been linked in recent years to the massive depletion of the ozone layer.

Change order: A permission request by a contractor or architect to make changes to an approved plan.

Cistern: A tank, above or below ground, used to hold fresh water.

Co-product: Anything left over from material processing that may be further processed and converted to usable materials.

Color temperature of light: The color appearance of a light, either cool or warm.

Combustion gases: Gases that are created through the process of burning, such as carbon monoxide. Gas appliances in a home may produce these gases, so proper ventilation is important.

Community: A group of several different species living in a defined habitat, which they share while also maintaining independence from each other.

Compact fluorescent lighting: A fluorescent light that is made to fit in an Edison light socket, used as an efficient alternative to incandescent lighting.

Composting: The process used to control decomposition of organic materials into a humus-like material that may be used as an organic fertilizer.

Composting toilet: A toilet that processes waste material into a material that may be used as a soil amendment.

Conduction: The transfer of heat through solid materials that are in contact with each other.

Conductor: Any substance or object able of transmitting heat, energy, or sound.

Constructed wetlands: Used to treat runoff and wastewater, a variety of systems designed using natural wetlands as a model.

Construction waste management: A term used to describe construction or demolition strategies that encourage recycling and reuse of materials.

Cooling/heating load: The amount of heat/cool air needed to offset a deficit/overage of the other.

Covenants: Worked into a deed, agreements that allow or disallow certain activities and usages on the deeded property.

CRI (color rendering index of light): When objects are illuminated by electric light, they will appear differently depending on the quality of the light. Rated on a scale of 1 to 100, objects will appear closer to their actual color with higher numbers and more distant with lower numbers.

Cross ventilation: The proper sizing and placements of doors, windows, and walls to cool a building by using natural breezes and air flow.

Critical zone: A location within a building that has numerous contaminant sources, thus requiring proper ventilation control to remain comfortable for occupants. Critical zones include: cafeterias, washrooms, auditoriums, smoking rooms, or any room where occupancy changes rapidly.

Cullet: Waste glass that has been crushed and returned for recycling.

Daylighting: The use of natural lighting for interiors through the placement of windows, skylights, and reflected light.

Degree days: A rough measure used to estimate the amount of heating required in a given area; the difference between the mean daily temperature and 65 degrees Fahrenheit.

Demand hot-water system: A hot-water heater designed to supply hot water instantaneously as opposed to storing preheated hot water in a tank. This system allows for the elimination of energy wasted to keep stored water warm, minimizes the amount of water wasted waiting for water to get warm, and helps to reduce conductive losses while traveling through pipes.

Design conditions: The interior and exterior environmental boundaries set for air conditioning and electrical design of a building.

Design temperatures: Temperatures used to base energy calculations on; calculations of extreme highs and lows established for different cities and different seasons.

Direct sunlight: The portion of daylight that arrives directly from the sun, without any diffusion.

Diurnal flux: The difference between daytime and nighttime temperatures, measured in degrees Fahrenheit. A diurnal flux of 25 degrees or more is considered an arid climate and is suitable for mass building construction.

Drip irrigation: A low-pressure, above-ground tube irrigation that constantly releases small amounts of water at the base of a plant.

Drought tolerance: The ability of a landscape plant to survive and function properly during drought conditions.

Earth sheltering (earth berming): Building a shelter below ground level, where soil temperatures are steadier and closer to the temperature desired, leading to 40-60 percent higher efficiency in heating and cooling systems.

Eave: The extension of the roof over the edge of a building, which serves to protect the structure's sides from the elements.

Ecology: From a biological standpoint, the study of relationships among living organisms and their surrounding environment. From a sociological standpoint, it is the interactions between human groups and their surroundings, also considering material resources and the consequential social and cultural patterns created.

Ecosystem: The delicate balances of a system of natural elements on which a habitat depends for survival.

Ecotourism: Tourism that partners with conservation efforts to preserve and protect the natural and cultural attractions of destination areas.

Edible landscaping: Using edible vegetation in a landscape design, such as fruit trees and fruit-bearing shrubs.

Efficiency: The ratio of energy input to useful energy output of a device.

Electricity: A flow of electrical power or a charge of energy created through friction, induction, or chemical change caused by the motion or presence of elementary charged particles.

Electronic ballast: A type of ballast for fluorescent lighting that can reduce flicker and noise, while also increasing efficiency.

Embodied energy: The energy needed to grow, harvest, extract, manufacture, refine, process,, pack, ship, install, and finally dispose of a particular object, product, or building material.

Emissivity: The ability of a material to convey far-infrared radiation through an air space. Materials with a low emissivity rating have a poor ability to do this and can be useful for blocking heat and controlling hot-air distribution in hot climates.

Encapsulation: A protective coating applied to asbestos-containing material in order to prevent the release of harmful fibers into the air.

End-use/least-cost: Focusing on the end user's needs, a decision-making tool to achieve the greatest benefits at the least cost in financial, social, as well as environmental terms.

Energy: The ability to do work; Btus and kilowatt-hours (kWhs) are common English units of energy measure.

Energy conservation: Efficiency of energy use, transmission, production, or distribution that creates a reduction in energy consumption while providing the same or higher levels of service.

Energy or water efficiency: The act of using less energy or water to perform the same basic tasks. A device is energy-efficient when it displays comparable or better quality of service while using a smaller amount of energy than other available technologies.

Engineered wood: Wood products that are reconstituted to result in various strengths and quality, producing a consistent product with less material.

Erosion: The wearing down of land surfaces caused by wind and water, the effects of which may be increased by land-clearing procedures related to farming, industrial or residential land development, logging, or road building.

ERV (energy-recovery ventilator): A type of mechanical equipment that uses an air-to-air heat exchanger in conjunction with ventilation systems to control air temperature and humidity levels in a building.

Evaporative cooling: Limited to arid climates, a cooling strategy that employs water evaporation into a hot, dry air stream in order to cool it.

Expanded polystyrene: A rigid insulation material, frequently made of recycled product and with CFC-free processing, made by heating pentane-saturated polystyrene pellets into various densities designed for different uses.

Feedstocks: Raw materials used to manufacture a product, such as the oil or gas required to produce a plastic.

Fenestration: A term that refers to any arrangement of openings in a building that admit daylight and any architectural elements that may affect light distribution.

Fluorescent lamp: A lamp that produces light by passing an electric arc between two tungsten cathodes in a tube filled with low-pressure mercury vapor and other gases.

Fly ash: Ash residue formed during high-temperature combustion processes, useful in the mixture of concrete.

Formaldehyde: Commonly used as an adhesive in wood products, this pungent-smelling, colorless material can be highly irritating if inhaled and is a probable human carcinogen.

Fossil fuels: Fuels, such as oil, coal, and natural gas, that take millions of years to form from the altered remains of once-living organisms.

Geotextiles: Fabrics engineered to be used in the soil for water and erosion control.

Geothermal/ground source heat pump: A heat pump that uses underground coils to collect and transmit heat to a building.

GIS (geographical information system): Used in the evaluation of probable building locations, landscaping, and land use, a technology that connects databases of information such as soil content, hydrology, and plant and animal habitats with mapping programs.

Glazing: Window glass, clear plastic films, or any type of transparent or translucent covering that protects from weather while still allowing light to pass through.

Global warming: As a result of greenhouse effects, a gradual increase of the earth's temperature over a long period of time.

GRAS (generally regarded as safe): A term used to describe products that have been used for many years without the occurrence of toxic side effects.

Graywater: Water that has been used for clothes washing, sinks, and showering and is suitable for reuse as subsurface irrigation in yards. Water from the kitchen sink and toilet is excluded from this category.

Green development: A development approach that is based on interconnected elements of environmental responsiveness, resource efficiency, as well as sensitivity to the existing community.

Greenhouse gas: A number of heat-trapping or radiatively active trace gases found in the earth's atmosphere that absorb infrared radiation, including water vapor, carbon dioxide, methane, ozone, CFCs, and nitrogen oxides.

Green roof: A green space integrated with or located on the roof of a building, consisting of a layer of living plants in a growing medium with a drainage system. It contributes to sustainable building strategies by reducing storm-water runoff, modulating temperatures in and around the building, and providing habitat for wildlife and functional open space for people.

Green wash: A false claim that a product or organization is environmentally sound, also known as faux green.

Habitat: A specific environment that organisms or biological populations rely on to live and grow.

HCFCs (hydrogen chlorofluorocarbons): A contributor to ozone depletion, but only one-twentieth as potent as CFCs.

Heat-island effect: A rise in ambient temperature in large, paved areas. Placed strategically, trees and landscaping elements can help reduce this effect while also substantially lowering cooling costs.

Heat pump: Used for heating and cooling, a mechanical device that moves heat from one location to another. Heat pumps draw from multiple sources of heat, including both air and water sources.

Heat-recovery ventilator: A device that, by forcing outgoing air past incoming air, can help retain temperature control and cut energy use for heating and cooling systems by 50-70 percent; also called an air-to-air heat exchanger.

HEPA (high-efficiency particulate-air) filter: A high-quality air filter, usually exceeding 98 percent atmospheric efficiency, typically used in clean rooms, surgeries, and in other special applications.

High-heeled truss: A roof-truss design with a space for insulation near the eaves. Other roof truss designs constrict the amount of space for insulation in this area.

High-mass construction: The passive building strategy of using heat-retaining materials, such as adobe and masonry, to help moderate diurnal temperature swings.

Household hazardous waste: Products used and disposed of in a residential setting that may be hazardous, including paints, stains, varnishes, solvents, pesticides, and any materials or chemicals that are corrosive, toxic, flammable, or explosive.

Humidistat: A device used to measure relative humidity.

Human health risk: The probability that a given exposure or series of exposures may have caused damage or could cause damage to an individual's health.

HVAC: Heating, ventilation, and air-conditioning system.

Hydronic heating: A system that heats space by using water circulated in a radiant floor or baseboard system or a fan-coil or convection system.

Impervious cover: A watertight barrier that covers the ground and does not allow water to seep into the soil below. Used correctly, it can prevent nonpoint source pollution at construction sites.

IAQ (indoor air quality): The health effects and cleanliness of the air in a building, highly influenced by the release of compounds into the space by various materials, microbial contaminants, and carbon-dioxide levels. Choice of building materials, cleaning procedures, and ventilation rates all highly determine IAQ.

Infill: A form of development that helps prevent urban sprawl and promotes economic revitalization by developing on empty lots of land already contained in an urban area, as opposed to on undeveloped lands outside of the established urban zone.

Infrastructure: Roads, highways, sewage, water, emergency services, parks and recreation, or any other service of facilities provided by a municipality or private source.

Insulation: Typically installed around living spaces to regulate and improve heating controls, a variety of materials, each with an R-value to define the resistance it has to heat flow. Materials with a higher R-value are more insulating and can be used effectively to slow the flow of heat.

Integrated design: Interactions between design, construction, and operations taken into consideration in an attempt to heal and protect damaged environments while reintroducing production of healthy food, clean water, and air into biologically healthy human communities.

Kilowatt (kW): A measure of electrical power, one kilowatt is equal to 1,000 watts.

Kilowatt-hour (kWh): A measurement of energy that uses the amount of power multiplied by the time of use to equal one unit.

Land stewardship: Managing land and resources in such a way as to promote sustainable and restorative actions.

Latent heat: The amount of heat required for a material to change phases (liquid to gas) without altering its temperature.

Latent load: Resulting from thermal energy, the cooling load released when air moisture changes from vapor to a liquid state. In humid, hot climates, cooling systems must have proper capacity to handle the latent load while still maintaining a comfortable climate for inhabitants.

Lead ventilation: The ventilation of a new, unoccupied building space to dilute contaminates from construction and HVAC systems to acceptable levels before occupants arrive.

Lease: A contract that allows a tenant to possess a domicile for a specified period of time while paying rent to a landlord.

Leichtlehm: Typically used in making walls, a mixture of straw and clay, moistened and molded between forms, that hardens into a strong material.

Life cycle: The sequential processes and interlinked stages of a product, beginning with its extraction and manufacturing and ending with recycling and waste-management operations.

Life-cycle assessment: The process of evaluating the cost of a product that considers all steps of a material's life cycle, including extraction and processing of raw materials, manufacturing, transportation, distribution, use, maintenance, reuse, recycling, and disposal.

Light: The visual perception of radiant energy.

Light construction: The construction of a building using materials of lower densities, reducing the capacity to store heat.

Light shelf: A daylighting strategy that bounces natural light off of a shelf below a window and onto a ceiling, bringing light deeper into the inside space.

Light-to-solar-gain ratio: The ratio of solar-heat gain to the ability of glazing to supply light.

Lignin: In wood, the naturally occurring polymer that binds the cellulose fibers together.

Linoleum: A natural and durable flooring material, primarily made from cork, that may also be used for other applications such as countertops.

Locally sourced materials: Obtaining materials from a defined radius to help lower impact by reducing transportation and energy usage while also supporting local economies.

Louvers: A set of baffles used to absorb unwanted light, shield light sources from view at certain angles, and allow for selective ventilation.

Low-emissivity windows: Glazing with special coatings used to allow most of the sun's light radiation through while preventing heat radiation from doing so.

Lumens: The amount of light given off by a light source.

Mass transit: The movement of people or goods from starting point to destination by use of public transportation systems such as bus, light rail, or subway.

MDF (medium-density fiberboard): A composite wood fiberboard, typically used in cabinetry and other interior applications, that may sometimes contain urea formaldehyde, contributing to poor indoor air quality.

Methane (CH$_4$): A colorless, odorless gas, nearly insoluble in water, that burns a pale, faintly luminous flame and produces water and carbon dioxide (or carbon monoxide if oxygen is not present).

Microclimate: The specific climatic conditions of a building site, affected by the site's geography, topography, vegetation, and proximity to bodies of water.

Mineral fibers: Used in insulation, glassy materials that have been melted and spun to create very fine fibers that are an inhalation hazard.

Mixed air: In an HVAC system, the mixture of outdoor air with return air, conditioned and filtered so it can be used for supply air.

Mixed-use development: A development strategy that integrates multiple revenue-raising uses into a single or multi-building plan, generally including housing, retail, and office space.

Mortgage: A written contract that requires real estate as collateral for the payment of a specified debt.

MSDs (material-safety data sheets): Documents required by OSHA to be provided by the manufacturer of products that are potentially hazardous. They contain information about potentially hazardous airborne contaminates, warnings, inspection tips, health effects, odor description, volatility, combustion contaminants, reactivity, and cleanup and disposal procedures.

Mulch: A layer of organic material (straw, wood chips, leaves, etc.) spread around plants to retain moisture, prevent growth of weeds, and enrich the soil.

National Fenestration Rating Council: A council that rates window models in a variety of areas such as light transmittance and energy efficiency.

Native vegetation: A plant that lives in a specific region naturally, not due to human cultivation or intervention.

Neotraditional planning: A type of planning, based on nineteenth-century American town plats, that aims to minimize automobile use and strengthen a sense of community by centering the plan on a town center and public open spaces.

New urbanism: An urban-planning movement that emphasizes inner-city revitalization and the reform of the American suburb community. Points emphasized include diversity of use and population; a focus on pedestrian-friendly access, and a well-defined public realm.

Nighttime ventilation: An energy-conserving building strategy that utilizes cool night-time air to flush the building and minimize the next day's cooling-energy load.

Nonpoint source pollution: Types of pollution that are difficult to link to one target source, typically pollutants of water.

Nonrenewable fuels: Fuels that are not easily manufactured or 'renewed,' Nonrenewable fuels, such as oil, natural gas, and coal, could one day be completely exhausted.

Nonrenewable resources: Resources that are limited and are used up much more quickly than they can be produced. These resources face exhaustion.

Occupancy sensor: A sensing device used to control lighting, ventilation, and heating settings based on whether the space is occupied or not.

Off-gas/out-gas: Emission of fumes into the air from new products such as new paint, carpeting, and various building materials. Many of these chemical compounds may be unpleasant and harmful to your health.

On-demand hot water: See demand hot water system.

Operating costs: All costs related to management of a property, including: maintenance, repairs, and operation of the property, utilities, insurance, and property taxes.

Organic matter: All natural materials of plant or animal origin.

Orientation: The alignment of a building to the directions of the compass and the sun.

OSHA (Occupational Safety and Health Administration): A federal agency created in 1971 for the purpose of preventing work-related injuries, illnesses, and deaths.

Outdoor air supply: Air that is brought into a building from outside.

Ozone (O_3): A molecule configured from three oxygen molecules. Ozone is poisonous at the earth's surface, but in the stratosphere the ozone layer protects the earth from harmful ultraviolet space radiation.

Particulate pollution: Pollution composed of tiny liquid or solid particles suspended in a water supply or the atmosphere.

Passive building design: Configurations of buildings that utilize natural and renewable resources, such as sunlight and cool air.

Passive cooling: A cooling strategy that combines the use of shaded windows, cooling summer breezes, and other factors to help reduce the cooling load of a building.

Passive solar system: Design of a building around the collection, storage, and distribution of solar resources.

Payback period: The amount of time required for a capital investment to pay back its initial investment, taking into account operating costs as well as profits earned.

Pedestrian pocket: A combination of housing, retail, and office space, placed within 1/4 mile of public transit, smaller in scale than planned unit developments or new towns.

Pedestrian scale: Urban designs that are oriented for pedestrians, making walking a safe, convenient, as well as interesting mode of travel.

Permeable: Allowing the passage of gases or fluids; relevant to moisture control and building materials.

Photocells: Light-sensitive cells that are used to activate switches at dawn and dusk.

Plasticizers: Chemicals that preserve the flexibility of soft plastics. Over time these agents off-gas, leaving the plastic brittle.

Porous paving: Paving materials that allow the infiltration of storm water, reducing the amount of runoff.

Post-consumer recycled content: Used materials that are separated from the waste stream for recycling.

Power: Expressed in watts (W), power is the rate at which energy is produced or consumed.

Pressure-treated wood: Wood chemically treated to prevent moisture and decay. The chemicals used can be a health hazard and must be treated and disposed of properly. Pressure-treated wood may create toxic fumes when burned.

Public transportation: Mass-transit systems, including buses and light rails. Sustainable building strategies place building sites near public transportation to promote commuting without using single-occupancy vehicles.

PVs (photovoltaics): Solid-state cells, typically made of silicon, that convert the sun's energy into electricity.

Radiant barrier: Material with a low emissivity rating, which is used to block the transfer of radiant heat across a space.

Radiant energy: Energy that radiates in all directions from its source in the form of electromagnetic waves.

Radiation: The transfer of heat through space by the straight-line passage of electro-magnetic waves from a warm object to a cooler one.

Radon gas: An odorless, colorless gas that is radioactive and naturally exists in soil or rocks. When concentrations of radon build up inside a building, it can become a serious health hazard.

Rafter: A structural part that generally supports roofing or decking.

Rammed earth: A high-mass mixture of water, earth, and cement, used to build walls.

Recycled material: Material that has been withheld from disposal and reintroduced as feed-stock, which is then manufactured into marketable end products.

Refrigerant: A substance used as the working fluid in a cooling system.

Relative humidity: The percentage of water vapor present in the air in relation to the capacity of air to hold water vapor before condensing to liquid form.

Relite: Translucent panels or windows placed above doors or high up on walls for the purpose of allowing light to penetrate deeper into a building.

Renewable energy: Energy sources that are virtually inexhaustible or regenerative, such as solar radiation, biomass, wind, water, and geothermal heat.

Renewable resources: Resources that, when managed properly, can be produced as quickly as they are consumed.

Renovation: Upgrading an existing building with new materials while maintaining the building's original appearance.

Resource conservation: Methods that attempt to protect, preserve, or renew natural resources in ways that provide the best economic and social benefits.

Restoration: Returning a landscape or structure to its original design.

Retrofit: Upgrading, replacing, or improving a structure or piece of equipment within an existing facility.

Reuse: The use of a product or component of municipal solid waste more than once while maintaining the same original form. Reuse of materials helps to reduce the strain on renewable and nonrenewable resources alike.

Ridge: The peak of a sloped roof.

Risk assessment: The evaluation of risks posed to human health and/or the environment due to actual or potential presence or exposure of specific pollutants.

Runoff: Water that flows over land instead of absorbing into the soil.

R-value: A unit used to rate the thermal resistance of insulating materials. A material with a high R-value has more insulating properties than one with a lower value.

Salvage: Materials that have been diverted from the waste stream for the purpose of being reused.

SBS (sick-building syndrome): A condition that is defined by the symptoms that people show when they are in an unhealthy building. The symptoms include dizziness, irritated eyes, headaches, nausea, throat irritation, and coughing and usually stop when the person exits the building.

Sealant: A compound used to seal or secure something to prevent the leakage of air or moisture.

Sediment basin: A depression made in the ground and placed so as to hold sediment and debris on site.

Shading coefficient: Measured in intervals between zero and one, the ratio of the solar heat gain of a given material to that of a 1/8-inch clear double-strength glass. A lower-rated window transmits less solar heat.

SIP (structural insulated panel): Manufactured wood panels that contain a core of polystyrene, making them more resistant to air infiltration.

Sisal: A fibrous product derived from the sisal plant, used for floor coverings.

Site assessment: Conducted in the planning stages of development, the assessment of site characteristics such as soil, hydrology, typography, wetlands, wind direction, solar orientation, and existing habitats as well as community connectedness.

Site development costs: All costs included in the preparation of land for development, including demolition of existing structures, site preparation, and on- and off-site improvements.

Skylight: A glazed roof aperture used to allow daylight into a building.

Sludge: The sediment extracted from wastewater.

Soffit: An eave with a closed underside.

Solar access: Laying out a building and its landscaping in order to allow maximum availability to the sun's energy.

Solar collector: Any device that uses the sun's energy to provide energy for uses that would usually be supplied by a non-renewable energy source. Photovoltaic panels are a solar collector.

Solar energy: Energy received from waves of electromagnetic radiation that originate from the sun.

Source reduction: The practice of designing, manufacturing, and purchasing materials that promote resource conservation while reducing the amount of toxic waste being released into the waste stream and the environment.

Spec house: A home constructed with the speculation of finding a buyer.

Specifications: As provided in conjunction with blueprints or plans, detailed instructions that give necessary instructions that are otherwise not included in a plan such as: building materials, dimensions, colors, or special construction techniques.

Sprawl: The extension of shopping centers, small industries, and residential areas outside the boundaries of a city.

Straw-bale construction: An alternative building method that utilizes bales of straw for wall construction. This method reduces the need for lumber products, reuses an agricultural waste product, and achieves high insulation values.

Stud: Vertical structural part used to frame walls, generally wood or metal.

Sulphur dioxide: A byproduct of coal combustion, a colorless, irritating gas that contributes to acid rain.

Sunshades: Blocking devices used to prevent unwanted solar gain.

Superwindows: With R-values of 4.5 or higher, double or triple-glazed windows that contain an interior layer of Mylar-coated film.

Supply air: Typically a conditioned mixture of return air and outdoor air; the entire amount of air supplied to a building or space for ventilation purposes.

Sustainable: The ability to meet the needs of present generations without compromising the needs of future generations. For a human community to remain sustainable, it must not compromise biodiversity, must reuse or recycle as many materials as possible, must not consume resources faster than they are renewed, and must rely mainly on local or regional resources.

Sustainably sourced materials: Materials that have been manufactured and handled in a way that emphasizes the appropriate and efficient use of natural resources.

Thermal break: A material with low conductance used to reduce the flow of heat between two elements.

Thermal bridging: An element with high conductivity that may compromise the insulating value of the building envelope.

Thermal chimney: An area of a building used to control and utilize hot-air currents to help stimulate the flow of fresh air into a building.

Thermal conductance: The ability of a material to conduct heat.

Thermal storage capacity: The capacity of a building material to internally store heat from the sun.

Tipping fees: Fees charged for dumping large amounts of waste into a landfill.

TND (traditional neighborhood development): One of the main forms of new urbanism, a development that has an identifiable edge and a center that includes public space as well as commercial enterprise, encompasses a variety of activities and housing types, uses a modified grid system of interconnected streets and blocks, and gives high priority to public spaces.

Topography: The configuration of the surface and physical features of a place.

Topsoil: The top layer of soil containing more organic material than other layers of soil.

Transit-oriented development: The development of a mixed-use community within a 2,000-foot average walking distance from a transit stop. Such developments combine a core commercial area as well as residential, retail, office, and open spaces into a public domain that is easily accessible by foot, bike, or transit.

Trombe wall: A wall of masonry, oriented facing south, that has a layer of glass spaced a few inches away. Rays from the sun pass through the glass and are transformed into heat on the surface of the wall; the heat either passes into the building interior or is thermosyphoned through vents to interior spaces.

Truth window (or wall): A section of window or wall that is cut away to show its internal components.

Ultraviolet radiation: Electromagnetic radiation that has wavelengths between 4 and 400 nanometers, usually comes from the sun, and can pose a health risk that may lead to cancer or cataracts.

Urea-formaldehyde foam insulation: An insulating material once commonly found in crawl spaces and attics, emissions from which have been found to be a health hazard.

U-value: The measure of heat conductivity in or out of a substance when the temperature on one side is one degree different from the other. U-values are used to measure the performance of a window assembly or glazing.

Vapor: The gaseous form of a compound that is usually in a liquid or solid form.

Vapor retarder/vapor barrier: A material used to stop or reduce the seepage of water.

Variance: A special allowance to use a specific property or structure in a way that would not normally be allowed because of existing zoning laws.

VOC (volatile organic compound): A chemical compound known to cause nausea, tremors, and headaches, and believed to cause long-lasting harm. VOCs are commonly emitted from oil-based paints, solvent-based finishes, and other construction materials.

Warm-edge technology: The placement of low-conductance spacers near the edge of insulated glazing to reduce heat transfer.

Wastewater: Water that has been used for residential, farming, or industrial processes and contains dissolved or particulate contaminants.

Water budget: An estimation of the water needs of a facility, taking into account fixture and appliance flow rates, occupancy needs, and landscaping needs.

Water harvesting: The collection of runoff and rainwater to be used for various tasks such as irrigation and water features.

Watershed: An area of land that collects draining water due to topographical features.

Watt (W): A measure of electrical power. One watt is equivalent to one joule of work per second.

Wetland: The transitional land between terrestrial and aquatic ecosystems that is covered by water for part of the year. Wetlands serve as natural flood protection, naturally improve water quality, and are also important for their habitat and diversity of species.

Whole-systems thinking: Taking into consideration all the interconnected systems in order to find, address, and solve multiple problems at once.

Wingwall: An outside wall, attached perpendicular to exterior walls, stimulating air movement near the window for ventilation purposes.

Work: The use of force through a distance. Power defines the rate at which work is done, and energy is stored work. The basic unit of work is the joule, which is defined as the amount of work done while exerting one newton (N) over one meter.

Xeriscaping: Landscape design that emphasizes water conservation by using drought-resistant and drought-tolerant plants.

Zoning: Local government rulings that serve to prevent conflicts in land use while maintaining structure in the development and regulation of privately owned land.

Time-Saving Tips and Tables for Builders

Grade	Ratio	Material needed for cu. yd.
Strong—watertight, exposed to weather and moderate wear	1:2 ¼:3	6 bags cement 14 cu. ft. sand (.52 yd^3) 18 cu. ft. stone (.67 yd^3)
Moderate—Strength, not exposed	1:2 ¾:4	5 bags cement 14 cu. ft. sand (.52 yd^3) 20 cu. ft. stone (.74 yd^3)
Economy—Massive areas, low strength	1:3:5	4½ bags cement 13 cu. ft. sand (.48 yd^3) 22 cu. ft. stone (.82 yd^3)

FIGURE A2-1   Concrete formulas.

Quantities of Concrete for Footings and Walls

Footing Size (")	Cubic Feet of concrete per LF	Cubic Yards of Concrete per 100 LF
6 x 12	0.50	1.9
8 x 12	0.67	2.5
8 x 16	0.89	3.3
10 x 12	0.83	3.1
10 x 16	1.11	4.1
10 x 18	1.25	4.6
12 x 12	1.00	3.7
12 x 24	2.00	7.4

Wall Thickness:

Thickness in inches	Square Feet	Thickness in inches	Square Feet
3	108	6	54
3½	93	6½	50
4	81	7	46
5	65	7½	43
5½	59	8	40

FIGURE A2-2   Quantities of concrete for footings and walls.

One Cubic Yard of Concrete Will Place

Thickness	Square Feet
2 inches	162
2½ inches	130
3 inches	108
3½ inches	93
4 inches	81
4½ inches	72
5 inches	65
5½ inches	59
6 inches	54
6½ inches	50
7 inches	46
7½ inches	43
8 inches	40
8½ inches	38
9 inches	36
9½ inches	34
10 inches	32.5
10½ inches	31
11 inches	29.5
11½ inches	28
12 inches	27
15 inches	21.5
18 inches	18
24 inches	13.5

FIGURE A2-3   Amount of coverage obtained from one cubic yard of concrete.

Wire size number		Nominal diameter, in.	Nominal weight, lb/ft	Area per width (in.^{2}/ft) for various spacings (in)						
Plain	Deformed			2	3	4	6	8	12	16
W45	D45	0.757	1.53	2.70	1.80	1.35	0.90	0.68	0.45	0.34
W31	D31	0.628	1.05	1.86	1.24	0.93	0.62	0.47	0.31	0.23
W20	D20	0.505	0.680	1.2	0.80	0.60	0.40	0.30	0.20	0.15
W18	D18	0.479	0.612	1.1	0.72	0.54	0.36	0.27	0.18	0.14
W16	D16	0.451	0.544	0.96	0.64	0.48	0.32	0.24	0.16	0.12
W14	D14	0.422	0.476	0.84	0.56	0.42	0.28	0.21	0.14	0.11
W12	D12	0.391	0.408	0.72	0.48	0.36	0.24	0.18	0.12	0.09
W11	D11	0.374	0.374	0.66	0.44	0.33	0.22	0.17	0.11	0.08
W10.5		0.366	0.357	0.63	0.42	0.32	0.21	0.16	0.11	0.08
W10	D10	0.357	0.340	0.60	0.40	0.30	0.20	0.15	0.10	0.08
W9.5		0.348	0.323	0.57	0.38	0.29	0.19	0.14	0.095	0.07
W9	D9	0.338	0.306	0.54	0.36	0.27	0.18	0.14	0.090	0.07
W8.5		0.329	0.289	0.51	0.34	0.26	0.17	0.13	0.085	0.06
W8	D8	0.319	0.272	0.48	0.32	0.24	0.16	0.12	0.080	0.06
W7.5		0.309	0.255	0.45	0.30	0.23	0.15	0.11	0.075	0.06
W7	D7	0.299	0.238	0.42	0.28	0.21	0.14	0.11	0.070	0.05
W6.5		0.288	0.221	0.39	0.26	0.20	0.13	0.097	0.065	0.05
W6	D6	0.276	0.204	0.36	0.24	0.18	0.12	0.090	0.060	0.05
W5.5		0.265	0.187	0.33	0.22	0.17	0.11	0.082	0.055	0.04
W5	D5	0.252	0.170	0.30	0.20	0.15	0.10	0.075	0.050	0.04
W4.5		0.239	0.153	0.27	0.18	0.14	0.090	0.067	0.045	0.03
W4	D4	0.226	0.136	0.24	0.16	0.12	0.080	0.060	0.040	0.03
W3.5		0.211	0.119	0.21	0.14	0.11	0.070	0.052	0.035	0.03
W3		0.195	0.102	0.18	0.12	0.090	0.060	0.045	0.030	0.02
W2.9		0.192	0.099	0.17	0.12	0.087	0.058	0.043	0.029	0.02
W2.5		0.178	0.085	0.15	0.10	0.075	0.050	0.037	0.025	0.02
W2.1		0.162	0.070	0.13	0.84	0.063	0.042	0.031	0.021	0.02
W2		0.160	0.068	0.12	0.080	0.060	0.040	0.030	0.020	0.02
W1.5		0.138	0.051	0.090	0.060	0.045	0.030	0.022	0.015	0.01
W1.4		0.134	0.048	0.084	0.056	0.042	0.028	0.021	0.014	0.01

(By permission, Concrete Reinforcing Steel Institute, Schramsburg, Illinois)

FIGURE A2-4   Welded wire mesh sizes.

Style designation (W = Plain, D = Deformed)	Steel area (in ²/ft)		Approximate weight (lb per 100 sq ft)
	Longitudinal	Transverse	
4 x 4-W1.4 x W1.4	0.042	0.042	31
4 x 4-W2.0 x W2.0	0.060	0.060	43
4 x 4-W2.9 x W2.9	0.087	0.087	62
4 x 4-W/D4 x W/D4	0.120	0:120	86
6 x 6-W1.4 x W1.4	0.028	0.028	21
6 x 6-W2.0 x W2.0	0.040	0.040	29
6 x 6-W2.9 x W2.9	0.058	0.058	42
6 x 6-W/D4 x W/D4	0.080	0.080	58
6 x 6-W/D4.7 x W/D4.7	0.094	0.094	68
6 x 6-W/D7.4 x W/D7.4	0.148	0.148	107
6 x 6-W/D7.5 x W/D7.5	0.150	0.150	109
6 x 6-W/D7.8 x W/D7.8	0.156	0.156	113
6 x 6-W/D8 x W/D8	0.160	0.160	116
6 x 6-W/D8.1 x W/D8.1	0.162	0.162	118
6 x 6-W/D8.3 x W/D8.3	0.166	0.166	120
12 x 12-W/D8.3 x W/D8.3	0.083	0.083	63
12 x 12-W/D8.8 x W/D8.8	0.088	0.088	67
12 x 12-W/D9.1 x W/D9.1	0.091	0.091	69
12 x 12-W/D9.4 x W/D9.4	0.094	0.094	71
12 x 12-W/D16 x W/D16	0.160	0.160	121
12 x 12-W/D16.6 x W/D16.6	0.166	0.166	126

*Many styles may be obtained in rolls.

FIGURE A2-5 Types of welded wire fabric.

BAR SIZE DESIGNATION	WEIGHT POUNDS PER FOOT	NOMINAL DIMENSIONS—ROUND SECTIONS		
		DIAMETER INCHES	CROSS-SECTIONAL AREA-SQ INCHES	PERIMETER INCHES
#3	.376	.375	.11	1.178
#4	.668	.500	.20	1.571
#5	1.043	.625	.31	1.963
#6	1.502	.750	.44	2.356
#7	2.044	.875	.60	2.749
#8	2.670	1.000	.79	3.142
#9	3.400	1.128	1.00	3.544
#10	4.303	1.270	1.27	3.990
#11	5.313	1.410	1.56	4.430
#14	7.650	1.693	2.25	5.320
#18	13.600	2.257	4.00	7.090

FIGURE A2-6 Rebar chart.

Cure time is a function of time, temperature, and type of cement used in the concrete mix.

The following cure times take these three factors into account.

At 50 degrees F (10 degrees C)-Measured in "days" required

Percentage design strength required	Type cement used in the mix		
	Type I	Type II	Type III
50%	6	9	3
65%	11	14	5
85%	21	28	16
95%	29	35	26

At 70 degrees F (21 degrees C) - Measured in "days" required

Percentage design strength required	Type cement used in the mix		
	Type I	Type II	Type III
50%	6	9	3
65%	11	14	5
85%	21	28	16
95%	29	35	26

FIGURE A2-7 Cure times.

					Modular brick sizes				
Unit		Dimensions (Inches)						Specified Dimensions Joint Thickness (Vertical)	
		Width-Height-Length W-H-L							
Modular	4	$2\frac{2}{3}$	8	$3\frac{3}{8}$	$2\frac{1}{4}$	$7\frac{5}{8}$	$\frac{3}{8}$	3C = 8 inches	
				$3\frac{1}{2}$	$2\frac{1}{4}$	$7\frac{1}{2}$	$\frac{1}{2}$		
Engineer Modular	4	$3\frac{1}{3}$	8	$3\frac{3}{8}$	$2\frac{3}{4}$	$7\frac{5}{8}$	$\frac{3}{8}$	5C = 16 inches	
				$3\frac{1}{2}$	$2\frac{13}{16}$	$7\frac{1}{2}$			
Closure Modular	4	4	8	$3\frac{5}{8}$	$3\frac{5}{8}$	$7\frac{5}{8}$	$\frac{3}{8}$	1C = 4 inches	
				$3\frac{1}{2}$	$3\frac{1}{2}$	$7\frac{1}{2}$	$\frac{1}{2}$		
Roman	4	2	12	$3\frac{5}{8}$	$1\frac{5}{8}$	$11\frac{5}{8}$		2C = 4 inches	
				$3\frac{1}{2}$	$1\frac{1}{2}$	$11\frac{1}{2}$	$\frac{1}{2}$		
Norman	4	$2\frac{2}{3}$	12	$3\frac{5}{8}$	$2\frac{1}{4}$	$11\frac{5}{8}$	$\frac{3}{8}$	3C = 8 inches	
				$3\frac{1}{2}$	$2\frac{1}{4}$	$11\frac{1}{2}$			
Engineer Norman	4	$3\frac{1}{2}$	12	$3\frac{5}{8}$	$2\frac{3}{4}$	$11\frac{5}{8}$	$\frac{3}{8}$	5C = 16 inches	
				$3\frac{1}{2}$	$2\frac{13}{16}$	$11\frac{1}{2}$	$\frac{1}{2}$		
Utility	4	4	12	$3\frac{5}{8}$	$3\frac{5}{8}$	$11\frac{5}{8}$	$\frac{3}{8}$	1C = 4 inches	
				$3\frac{1}{2}$	$3\frac{1}{2}$	$11\frac{1}{2}$	$\frac{1}{2}$		

Metric equivalents for modular brick sizes

$4 \times 2\frac{2}{3} \times 8 = 10.16cm \times 5.95cm \times 20.32cm$
$3\frac{3}{8} \times 2\frac{1}{4} \times 7\frac{5}{8} = 8.57cm \times 5.63 \times 18.1cm$
$3\frac{1}{2} \times 2\frac{1}{4} \times 7\frac{1}{2} = 8.75cm \times 5.63cm \times 18.75cm$

$4 \times 3 - \frac{1}{3} \times 8 = 10.16cm \times 8.33cm \times 20cm$
$3\frac{3}{8} \times 2\frac{3}{4} \times 7\frac{5}{8} = 8.57cm \times 5.63cm \times 18.1cm$
$3\frac{1}{2} \times 2\frac{13}{16} \times 7\frac{1}{2} = 8.75cm \times 5.474cm \times 18.75cm$

$4 \times 4 \times 8 = 10.16cm \times 10.16cm \times 20cm$
$3\frac{5}{8} \times 3\frac{5}{8} \times 7 - \frac{5}{8} = 9.2cm \times 9.2cm \times 18.1cm$
$3\frac{1}{2} \times 3\frac{1}{2} \times 7\frac{1}{2} = 8.75cm \times 8.75cm \times 18.75cm$

$4 \times 2 \times 12 = 10.16cm \times 5cm \times 30cm$
$3\frac{5}{8} \times 1\frac{5}{8} \times 11\frac{5}{8} = 9.2cm \times 4.127cm \times 29.52cm$
$3\frac{1}{2} \times 1\frac{1}{2} \times 11\frac{1}{2} = 8.75cm \times 3.75cm \times 28.75cm$

$4 \times 2\frac{2}{3} \times 12 = 10.16cm \times 5.95cm \times 30cm$
$3\frac{5}{8} \times 2\frac{1}{4} \times 11 - \frac{5}{8} = 9.2cm \times 5.63cm \times 29.52cm$
$3\frac{1}{2} \times 2\frac{1}{4} \times 11\frac{1}{2} = 8.75cm \times 5.63cm \times 28.75cm$

$4 \times 3\frac{1}{2} \times 12 = 10.16cm \times 8.75cm \times 30cm$
$3\frac{5}{8} \times 2\frac{3}{4} \times 11\frac{5}{8} = 9.2cm \times 5.63cm \times 29.52cm$
$3\frac{1}{2} \times 2 - \frac{13}{16} \times 11\frac{1}{2} = 8.75cm \times 5.47cm \times 28.75cm$

$4 \times 4 \times 12 = 10.16cm \times 10.16cm \times 30cm$
$3\frac{5}{8} \times 3\frac{5}{8} \times 11\frac{5}{8} = 9.2cm \times 9.2cm \times 29.52cm$
$3\frac{1}{2} \times 3\frac{1}{2} \times 11\frac{1}{2} = 8.75cm \times 8.75cm \times 28.75cm$

Note: 2C, 5C etc refers to number of courses and "inches" refers to height of that coursing.

FIGURE A2-8   Modular brick sizes.

Other modular brick sizes				
Nominal Size	Specified Dimensions (Inches)		Joint Thickness	Vertical Coursing
W H L	W H L			
4 6 8	3½ 5½ ½		½	2C = 12 inches
4 8 8	3½ 7½ 7½		½	1 C = 8 inches
6 3½ 12	5½ 2³⁄₁₆ 11½		½	5 C = 16 inches
6 4 12	5½ 3½ 11½		½	1C = 4 inches
8 4 12	7½ 3½ 11½		½	1 C = 4 inches
8 4 16	7½ 3½ 15½		½	1C = 4 inches

Metric equivalents for other modular brick sizes

4 x 6 x 8 = 10.16cm x 15cm x20cm
3½ x 5½ x 7½ =8.75cm x 13.75cm x 18.75cm

4 x 8 x 8 = 10.16cmx20cmx20cm
3½ x 7½ x 7½ = 8.75cm x 18.75cm x 18.75cm

6 x 3 1/2 x 12 = 15cm x 8.75cm x 30cm
5½ x 2³⁄₁₆ x 11½ =13.75cm x 5.47cmx28.75cm

6 x 4 x 12 = 15cm x 10.16cm x 30cm
5½ x 3½ x 11½ = 13.75cm x 8.75cm x 28.75cm

8 x 4 x 12 = 20cm x 10.16cm x 30cm
7½ x 3½ x 11½ = 18.75cm x 8.75cm x 28.75 cm

8 x 4x 16 = 20cm x 10.16 cm x 40 cm
7½ x 3½ x 15½ = 18.75 x 8.75cm x 38.75 cm

Other conversions for the previous page; ½" = 1.25cm, 12" = 30.48cm, 4" = 10.16cm, 8" = 20.32cm

Note: Specified dimensions may vary somewhat from manufacturer to manufacturer.

FIGURE A2-9 Other modular brick sizes.

Non-modular brick coursing				
Unit	Specified Dimensions (Inches)		Joint Thickness	Vertical Coursing
	W H L			
Standard	3⅝ 2¼ 8		3/8	3C = 8 inches
	3½ 2¼ 8		½	
Engineer Standard	3⅝ 2¾ 8		⅝	5C = 16 inches
	3½ 2¹³⁄₁₆ 8		½	
Closure Standard	3⅝ 3⅝ 8		3/8	1C = 4 inches
	3½ 3½ 8		½	
King	3 2¾ 9⅝		⅜	5C = 16 inches
	3 3⅝ 9¾		⅜	
Queen	3 2¾ 9¾		⅜	5C= 16 inches
Metric equivalents for non-modular brick coursing				

3⅝ x 2¼ x 8 = 9.20cm x 5.63cm x 20.32 cm   ⅜" = 9.5 mm
3½ x 2¼ x 8 = 8.75cm x 5.63cm x 20.32cm   ½" = 1.25 cm

3⅝ x 2¾ x 8 = 9.20cm x 6.98cm x 20.32cm
3½ x 2¹³⁄₁₆ x 8 = 8.75cm x 5.47cm x 20.32cm

3⅝ x 3⅝ x 8 = 9.20 cm x 9.20cm x 20.32cm
3½ x 3½ x 8 = 8.75cm x 8.75cm x 20.32

3 x 2¾ x 9⅝ = 7.5cm x 6.98cm x 24.08cm
3 x 3⅝ x 9¾ = 7.5cm x 9.20 x 24.37 cm

3 x 2¾ x 9¾ = 7.5 cm x 6.87cm x 24.37 cm

3 x 2⅝ x 8⅝ = 7.5cm x 6.58cm x 20.16 cm

Note: Specified dimensions may vary within this range from manufacturer to manufacturer.

FIGURE A2-10   Non-modular brck coursing.

Beam size	Maximum distance between support post
4-x-6	6 feet
4-x-8	8 feet
4-x-10	10 feet
4-x-12	12 feet

FIGURE A2-11   Common beam spans for decks.

Beam size	Maximum span
2-x-6	8 feet
2-x-8	10 feet
2-x-10	13 feet

FIGURE A2-12   Common joint spans for decks where joists are installed 16 inches on center.

Beam size	Maximum span
2-x-6	5 feet
2-x-8	7 feet
2-x-10	8 feet

FIGURE A2-13   Common joist spans for decks where joists are installed 24 inches on center.

Types of Plywood with Typical Uses

Softwood veneer	Cross laminated plies or veneers—Sheathing, general construction and industrial use, etc.
Hardwood veneer	Cross laminated plies with hardwood face and back veneer—Furniture and cabinet work, etc.
Lumbercore plywood	Two face veneers and 2 crossband plies with an inner core of lumber strips—Desk and table tops, etc.
Medium-density overlay (MDO)	Exterior plywood with resin and fiber veneer—Signs, soffits, etc.
High-density overlay (HDO)	Tougher than MDO—Concrete forms, workbench tops, etc.
Plywood siding	T-111 and other textures used as one step sheathing and siding where codes allow.

FIGURE A2-14 Types of plywood with typical uses.

Reconstituted Wood
Panels with Typical Uses

Particleboard	Wood particles and resin. a. Industrial grade—Cabinets and counter tops under plastic laminates. b. Underlayment—Installed over subfloor under tile or carpet.
Wafer board	Wood wafers and resin. Inexpensive sheathing, craft projects, etc.
Oriented Strand Board (OSB)	Thin wood strands oriented at right angles with phenolic resin. Same uses as above.
Hardboard	Wood fiber mat compressed into stiff, hard sheets. a. Service—Light weight—Cabinet backs, etc. b. Standard—Stronger with better finish quality. c. Tempered—Stiffer and harder for exterior use.
Fiberboard (Grayboard)	Molded wood fibers—Underlayment, sound deadening panels.
Composite plywood	A core of particleboard with face and back veneers glued directly to it.

FIGURE A2-15   Reconstructed wood panels with typical uses.

Size	Lgth (in.)	Diam (in.)	Remarks	Where used
2d	1	.072	Small head	Finish work, shop work.
2d	1	.072	Large flathead	Small timber, wood shingles, lathes.
3d	1¼	.08	Small head	Finish work, shop work.
3d	1¼	.08	Large flathead	Small timber, wood shingles, lathes.
4d	1½	.098	Small head	Finish work, shop work.
4d	1½	.098	Large flathead	Small timber, lathes, shop work.
5d	1¾	.098	Small head	Finish work, shop work.
5d	1¾	.098	Large flathead	Small timber, lathes, shop work.
6d	2	.113	Small head	Finish work, casing, stops, etc., shop work.
6d	2	.113	Large flathead	Small timber, siding, sheathing, etc., shop work.
7d	2¼	.113	Small head	Casing, base, ceiling, stops, etc.
7d	2¼	.113	Large flathead	Sheathing, siding, subflooring, light framing.
8d	2½	.131	Small head	Casing, base, ceiling, wainscot, etc., shop work.
8d	2½	.131	Large flathead	Sheathing, siding, subflooring, light framing, shop work.
8d	1¼	.131	Extra-large flathead	Roll roofing, composition shingles.
9d	2¾	.131	Small head	Casing, base, ceiling, etc.
9d	2¾	.131	Large flathead	Sheathing, siding, subflooring, framing, shop work.
10d	3	.148	Small head	Casing, base, ceiling, etc., shop work.
10d	3	.148	Large flathead	Sheathing, siding, subflooring, framing, shop work.
12d	3¼	.148	Large flathead	Sheathing, subflooring, framing.
16d	3½	.162	Large flathead	Framing, bridges, etc.
20d	4	.192	Large flathead	Framing, bridges, etc.
30d	4½	.207	Large flathead	Heavy framing, bridges, etc.
40d	5	.225	Large flathead	Heavy framing, bridges, etc.
50d	5½	.244	Large flathead	Extra-heavy framing, bridges, etc.
60d	6	.262	Large flathead	Extra-heavy framing, bridges, etc.

[1] This chart applies to wire nails, although it may be used to determine the length of cut nails.

FIGURE A2-16   Recommended uses of nails (chart courtesy of USDA Forest Service).

Nail Sizes and Number Per Pound

Penny size "d"	Length	Approximate number per pound, Common	Approximate number per pound, Box	Approximate number per pound, Finish
2	1"	875	1000	1300
3	1¼"	575	650	850
4	1½"	315	450	600
5	1¾"	265	400	500
6	2"	190	225	300
7	2¼"	160		
8	2½"	105	140	200
9	2¾"	90		
10	3"	70	90	120
12	3¼"	60	85	110
16	3½"	50	70	90
20	4"	30	50	60
30	4½"	25		
40	5"	20		
50	5½"	15		
60	6"	10		

NOTE: Aluminum and c.c. nails are slightly smaller than other nails of the same penny size.

FIGURE A2-17   Nail sizes and the approximate number of nails per pound.

Screw Lengths and Available Gauge Numbers

Length	Gauge numbers	Length	Gauge numbers
¼"	0 to 3	1¾"	8 to 20
⅜"	2 to 7	2"	8 to 20
½"	2 to 8	2¼"	9 to 20
⅝"	3 to 10	2½"	12 to 20
¾"	4 to 11	2¾"	14 to 20
⅞"	6 to 12	3"	16 to 20
1"	6 to 14	3½"	18 to 20
1¼"	7 to 16	4"	18 to 20
1½"	6 to 18		

FIGURE A2-18   Screw lengths and available gauge numbers.

Fractions to Decimals

Fractions	Decimal equivalent	Fractions	Decimal equivalent
¹⁄₁₆	0.0625	⁹⁄₁₆	0.5625
⅛	0.1250	⅝	0.6250
³⁄₁₆	0.1875	¹¹⁄₁₆	0.6875
¼	0.2500	¾	0.7500
⁵⁄₁₆	0.3125	¹³⁄₁₆	0.8125
⅜	0.3750	⅞	0.8750
⁷⁄₁₆	0.4375	¹⁵⁄₁₆	0.9375
½	0.5000	1	1.000

FIGURE A2-19   Converting fractions to decimals.

Inches to Millimetres

Inches	Millimetres	Inches	Millimetres
1	25.4	11	279.4
2	50.8	12	304.8
3	76.2	13	330.2
4	101.6	14	355.6
5	127.0	15	381.0
6	152.4	16	406.4
7	177.8	17	431.8
8	203.2	18	457.2
9	228.6	19	482.6
10	254.0	20	508.0

FIGURE A2-20   Converting inches to millimeters.

Square Inches to Approximate Square Centimeters

Square inches	Square centimeters	Square inches	Square centimeters
1	6.5	8	52.0
2	13.0	9	58.5
3	19.5	10	65.0
4	26.0	25	162.5
5	32.5	50	325.0
6	39.0	100	650.0
7	45.5		

FIGURE A2-21 Coverting square inches to square centimeters.

Square Feet to Approximate Square Meters

Square feet	Square meters	Square feet	Square meters
1	0.925	8	0.7400
2	0.1850	9	0.8325
3	0.2775	10	0.9250
4	0.3700	25	2.315
5	0.4650	50	4.65
6	0.5550	100	9.25
7	0.6475		

FIGURE A2-22 Converting square feet to square meters.

Formulas

Circle

Circumference = diameter × 3.1416

Circumference = radius × 6.2832

Diameter = radius × 2

Diameter = square root of; (area ÷ 0.7854)

Diameter = square root of area × 1.1283

Diameter = circumference × 0.31831

Radius = diameter ÷ 2

Radius = circumference × 0.15915

Radius = square root of area × 0.56419

Area = diameter × diameter × 0.7854

Area = half of the circumference × half of the diameter

Area = square of the circumference × 0.0796

Arc length = degrees × radius × 0.01745

Degrees of arc = length ÷ (radius × 0.01745)

Radius of arc = length ÷ (degrees × 0.01745)

Side of equal square = diameter × 0.8862

Side of inscribed square = diameter × 0.7071

Area of sector = area of circle × degrees of arc ÷ 360

Cone

Area of surface = one half of circumference of base × slant height + area of base.

Volume = diameter × diameter × 0.7854 × one-third of the altitude.

FIGURE A2-23   Formula functions. (continued)

Cube

Volume = width × height × length

Cylinder

Area of surface = diameter × 3.1416 × length + area of the two bases
Area of base = diameter × diameter × 0.7854
Area of base = volume ÷ length
Length = volume ÷ area of base
Volume = length × area of base
Capacity in gallons = volume in inches ÷ 231
Capacity of gallons = diameter × diameter × length × 0.0034
Capacity in gallons = volume in feet × 7.48

Ellipse

Area = short diameter × long diameter × 0.7854

Hexagon

Area = width of side × 2.598 × width of side

Parallelogram

Area = base × distance between the two parallel sides

Pyramid

Area = ½ perimeter of base × slant height + area of base
Volume = area of base × ⅓ of the altitude

Rectangle

Area = length × width

Rectangular prism

Volume = width × height × length

FIGURE A2-23 Formula functions. (continued)

Sphere

Area of surface = diameter × diameter × 3.1416

Side of inscribed cube = radius × 1.547

Volume = diameter × diameter × diameter × 0.5236

Square

Area = length × width

Triangle

Area = one-half of height times base

Trapezoid

Area = one-half of the sum of the parallel sides × the height

FIGURE A2-23 Formula functions.

Board Lumber Measure

Nominal size	Actual size	Board feet per linear foot	Linear feet per 1000 board feet
1 × 2	¾ × 1½	⅙ (0.167)	6000
1 × 3	¾ × 2½	¼ (0.250)	4000
1 × 4	¾ × 3½	⅓ (0.333)	3000
1 × 6	¾ × 5½	½ (0.500)	2000
1 × 8	¾ × 7¼	⅔ (0.666)	1500
1 × 10	¾ × 9¼	⅚ (0.833)	1200
1 × 12	¾ × 11¼	1 (1.0)	1000

FIGURE A2-24   Board lumber conversions.

Dimensional Lumber Board Measure

Nominal size	Actual size	Board feet per linear foot	Linear feet per 1000 board feet
2 × 2	1½ × 1½	⅓ (0.333)	3000
2 × 3	1½ × 2½	½ (0.500)	2000
2 × 4	1½ × 3½	⅔ (0.666)	1500
2 × 6	1½ × 5½	1 (1.0)	1000
2 × 8	1½ × 7¼	1⅓ (1.333)	750
2 × 10	1½ × 9¼	1⅔ (1.666)	600
2 × 12	1½ × 11¼	2 (2.0)	500

FIGURE A2-25   Dimensional lumber board measures.

Chord size: 2-×-4 top and 2-×-4 bottom	
Pitch	**Span (feet)**
2/12	22
3/12	29
4/12	33
5/12	35
6/12	37

Typical truss spans (55 PSF with 15% duration factor).

Chord size: 2-×-6 top and 2-×-6 bottom	
Pitch	**Span (feet)**
2/12	32
3/12	51
4/12	56
5/12	60
6/12	62

Monopitch truss spans (55 PSF with 33% duration factor).

Chord size: 2-×-6 top and 2-×-4 bottom	
Pitch	**Span (feet)**
2/12	28
3/12	39
4/12	46
5/12	53
6/12	57

Typical truss spans (47 PSF with 33% duration factor).

FIGURE A2-26 Truss spans.

Chord size: 2-×-4 top and 2-×-4 bottom	
Pitch	**Span (feet)**
2/12	25
3/12	33
4/12	37
5/12	40
6/12	41

Monopitch truss spans (55 PSF with 33% duration factor).

Chord size: 2-×-6 top and 2-×-4 bottom	
Pitch	**Span (feet)**
2/12	23
3/12	31
4/12	39
5/12	45
6/12	51

Monopitch truss spans (55 PSF with 15% duration factor).

Chord size: 2-×-4 top and 2-×-4 bottom	
Pitch	**Span (feet)**
2/12	22
3/12	30
4/12	33
5/12	35
6/12	37

Monopitch truss spans (55 PSF with 15% duration factor).

FIGURE A2-27 Truss spans.

Chord size: 2-×-6 top and 2-×-6 bottom	
Pitch	**Span (feet)**
2/12	34
3/12	45
4/12	50
5/12	53
6/12	56

Monopitch truss spans (55 PSF with 15% duration factor).

Chord size: 2-×-6 top and 2-×-4 bottom	
Pitch	**Span (feet)**
2/12	25
3/12	35
4/12	42
5/12	48
6/12	53

Monopitch truss spans (55 PSF with 33% duration factor).

Chord size: 2-×-6 top and 2-×-6 bottom	
Pitch	**Span (feet)**
2/12	44
3/12	57
4/12	63
5/12	67
6/12	68

Monopitch truss spans (47 PSF with 33% duration factor).

FIGURE A2-28   Truss spans.

Chord size: 2-×-4 top and 2-×-4 bottom	
Pitch	**Span (feet)**
2/12	28
3/12	38
4/12	42
5/12	44
6/12	45

Monopitch truss spans (47 PSF with 33% duration factor).

Chord size: 2-×-6 top and 2-×-4 bottom	
Pitch	**Span (feet)**
2/12	28
3/12	38
4/12	46
5/12	52
6/12	57

Typical truss spans (47 PSF with 33% duration factor).

Chord size: 2-×-4 top and 2-×-4 bottom	
Pitch	**Span (feet)**
2/12	28
3/12	37
4/12	41
5/12	44
6/12	44

Typical truss spans (47 PSF with 33% duration factor).

FIGURE A2-29   Truss spans.

Chord size: 2-×-6 top and 2-×-4 bottom	
Pitch	Span (feet)
2/12	25
3/12	34
4/12	42
5/12	48
6/12	53

Typical truss spans (55 PSF with 33% duration factor).

Chord size: 2-×-4 top and 2-×-4 bottom	
Pitch	Span (feet)
2/12	25
3/12	33
4/12	37
5/12	39
6/12	41

Typical truss spans (55 PSF with 33% duration factor).

Chord size: 2-×-6 top and 2-×-6 bottom	
Pitch	Span (feet)
2/12	39
3/12	50
4/12	55
5/12	59
6/12	62

Typical truss spans (55 PSF with 33% duration factor).

FIGURE A2-30 Truss spans.

Chord size: 2-×-6 top and 2-×-4 bottom	
Pitch	**Span (feet)**
2/12	23
3/12	31
4/12	39
5/12	45
6/12	51

Typical truss spans (55 PSF with 15% duration factor).

Chord size: 2-×-6 top and 2-×-6 bottom	
Pitch	**Span (feet)**
2/12	34
3/12	44
4/12	49
5/12	53
6/12	55

Typical truss spans (55 PSF with 15% duration factor).

Chord size: 2-×-4 top and 2-×-4 bottom		
Top chord pitch	**Bottom chord pitch**	**Span (feet)**
6/12	2/12	36
6/12	3/12	31
6/12	4/12	24

Scissor truss spans (55 PSF with 33% duration factor).

FIGURE A2-31   Truss spans.

Chord size: 2-×-6 top and 2-×-6 bottom		
Top chord pitch	**Bottom chord pitch**	**Span (feet)**
6/12	2/12	54
6/12	3/12	48
6/12	4/12	36

Scissor truss spans (55 PSF with 33% duration factor).

Chord size: 2-×-6 top and 2-×-4 bottom		
Top chord pitch	**Bottom chord pitch**	**Span (feet)**
6/12	2/12	38
6/12	3/12	30
6/12	4/12	22

Scissor truss spans (55 PSF with 15% duration factor).

Chord size: 2-×-4 top and 2-×-4 bottom		
Top chord pitch	**Bottom chord pitch**	**Span (feet)**
6/12	2/12	32
6/12	3/12	28
6/12	4/12	21

Scissor truss spans (55 PSF with 15% duration factor).

Chord size: 2-×-6 top and 2-×-6 bottom		
Top chord pitch	**Bottom chord pitch**	**Span (feet)**
6/12	2/12	48
6/12	3/12	42
6/12	4/12	32

Scissor truss spans (55 PSF with 15% duration factor).

FIGURE A2-32   Truss spans.

Chord size: 2-×-6 top and 2-×-4 bottom		
Top chord pitch	Bottom chord pitch	Span (feet)
6/12	2/12	42
6/12	3/12	34
6/12	4/12	24

Scissor truss spans (55 PSF with 33% duration factor).

Chord size: 2-×-6 top and 2-×-6 bottom		
Top chord pitch	Bottom chord pitch	Span (feet)
6/12	2/12	61
6/12	3/12	54
6/12	4/12	42

Scissor truss spans (47 PSF with 33% duration factor).

Chord size: 2-×-6 top and 2-×-4 bottom		
Top chord pitch	Bottom chord pitch	Span (feet)
6/12	2/12	46
6/12	3/12	38
6/12	4/12	27

Scissor truss spans (47 PSF with 33% duration factor).

Chord size: 2-×-4 top and 2-×-4 bottom		
Top chord pitch	Bottom chord pitch	Span (feet)
6/12	2/12	40
6/12	3/12	35
6/12	4/12	27

Scissor truss spans (47 PSF with 33% duration factor).

FIGURE A2-33   Truss spans.

Defining Roof Slopes and other Types of Slopes

Percent Slope	Inch/Ft	Ratio	Degrees from Horizontal
1%	1/8	1 in 100	—
2%	1/4	1 in 50	—
3%	3/8	—	—
4%	1/2	1 in 25	—
5%	5/8	1 in 20	3
6%	3/4	—	—
7%	7/8	—	—
8%	approx. 1	approx. 1 in 12	—
9%	1 1/8	—	—
10%	1 1/4	1 in 10	6
11%	1 3/8	approx. 1 in 9	—
12%	1 1/2	—	—
13%	1 5/8	—	—
14%	1 3/4	—	—
15%			8.5
16%	1 7/8	—	—
17%	2	approx. 2 in 12	—
18%	2 1/8	—	—
19%	2 1/4	—	—
20%	2 3/8	1 in 5	11.5
25%	3	3 in 12	14
30%	3.6	1 in 3.3	17
35%	4.2	approx. 4 in 12	19.25
40%	4.8	approx. 5 in 12	21.5
45%	5.4	1 in 22	24
50%	6	6 in 12	26.5
55%	6 5/8	1 in 1.8	28.5
60%	7 1/4	approx. 7 in 12	31
65%	7 3/4	1 in 1 1/2	33
70%	8 1/8	1 in 1.4	35
75%	9	1 in 1.3	36.75
100%	12	1 in 1	45

FIGURE A2-34   Defining roof slope as a percentage and inches per foot.

Slate Thickness	Sloping Roof With 3'' (76mm) Lap (approx. Pounds Per Square [kg/m²])
3/16'' to 1/4'' (4mm to 6mm)	700 to 1,000 lbs/sq (3,417 to 4,882 kg/m²)
3/8'' (9mm)	1,500 lbs/sq (7,323 kg/m²)
1/2'' (13mm)	2,000 lbs/sq (9,764 kg/m²)
3/4'' (19mm)	3,000 lbs/sq (14,646 kg/m²)
1'' (25mm)	4,000 lbs/sq (19,528 kg/m²)
1 1/4'' (32mm)	5,000 lbs/sq (24,410 kg/m²)
1 1/2'' (38mm)	6,000 lbs/sq (29,292 kg/m²)
1 3/4'' (44mm)	7,000 lbs/sq (34,174 kg/m²)
2'' (51mm)	8,000 lbs/sq (39,056 kg/m²)

FIGURE A2-35   Slate shingle size, weight, and exposures.

SCHEDULE FOR STANDARD 3/16"(5mm) THICK SLATE

SIZE OF SLATE (L x W) (IN.)	SIZE OF SLATE (L x W) (MM)	SLATES PER SQUARE	EXPOSURE WITH 3"(76mm) Lap (IN.)	EXPOSURE WITH 3"(76mm) Lap (MM)	SIZE OF SLATE (L x W) (IN.)	SIZE OF SLATE (L x W) (MM)	SLATES PER SQUARE	EXPOSURE WITH 3"(76mm) Lap (IN.)	EXPOSURE WITH 3"(76mm) Lap (MM)
26 x 14	660 x 356	89	11 1/2	292	16 x 14	406 x 356	160	6 1/2	165
					16 x 12	406 x 305	184	6 1/2	165
24 x 16	610 x 406	86	10 1/2	267	16 x 11	406 x 279	201	6 1/2	165
24 x 14	610 x 356	98	10 1/2	267	16 x 10	406 x 254	222	6 1/2	165
24 x 13	610 x 330	106	10 1/2	267	16 x 9	406 x 229	246	6 1/2	165
24 x 12	610 x 305	114	10 1/2	267	16 x 8	406 x 203	277	6 1/2	165
24 x 11	610 x 279	138	10 1/2	267					
					14 x 12	356 x 305	218	5 1/2	140
22 x 14	559 x 356	108	9 1/2	241	14 x 11	356 x 279	238	5 1/2	140
22 x 13	559 x 330	117	9 1/2	241	14 x 10	356 x 254	261	5 1/2	140
22 x 12	559 x 305	126	9 1/2	241	14 x 9	356 x 229	291	5 1/2	140
22 x 11	559 x 279	138	9 1/2	241	14 x 8	356 x 203	327	5 1/2	140
22 x 10	559 x 254	152	9 1/2	241	14 x 7	356 x 178	374	5 1/2	140
20 x 14	508 x 356	121	8 1/2	216	12 x 10	305 x 254	320	4 1/2	114
20 x 13	508 x 330	132	8 1/2	216	12 x 9	305 x 229	355	4 1/2	114
20 x 12	508 x 305	141	8 1/2	216	12 x 8	305 x 203	400	4 1/2	114
20 x 11	508 x 279	154	8 1/2	216	12 x 7	305 x 178	457	4 1/2	114
20 x 10	508 x 254	170	8 1/2	216	12 x 6	305 x 152	533	4 1/2	114
20 x 9	508 x 229	189	8 1/2	216					
					11 x 8	279 x 203	450	4	102
18 x 14	457 x 356	137	7 1/2	191	11 x 7	279 x 178	515	4	102
18 x 13	457 x 330	148	7 1/2	191					
18 x 12	457 x 305	160	7 1/2	191	10 x 8	254 x 203	515	3 1/2	89
18 x 11	457 x 279	175	7 1/2	191	10 x 7	254 x 178	588	3 1/2	89
18 x 10	457 x 254	192	7 1/2	191	10 x 6	254 x 152	686	3 1/2	89
18 x 9	457 x 229	213	7 1/2	191					

FIGURE A2-36   Schedule for standard slate.

SHINGLE COVERAGE

LENGTH OF NO. 1 SHINGLES IN INCHES (mm)	APPROXIMATE COVERAGE IN SQ. FT. (m²) OF ONE SQUARE (4 BUNDLES) OF SHINGLES BASED ON FOLLOWING WEATHER EXPOSURES IN INCHES (mm)								
	3 1/2" (89)	4" (102)	4 1/2" (115)	5" (127)	5 1/2" (140)	6" (152)	6 1/2" (165)	7" (178)	7 1/2" (191)
16" (406)	70 (6.50)	80 (7.43)	90 (8.36)	100* (9.29)	—	—	—	—	—
18" (457)	—	72.5 (6.74)	81.5 (7.57)	90.5 (8.40)	100* (9.29)	—	—	—	—
24" (610)	—	—	—	—	73.5 (6.83)	80 (7.43)	86.5 (8.04)	93 (8.64)	100* (9.29)

* MAXIMUM EXPOSURE RECOMMENDED FOR ROOFS

NO.1

FIGURE A2-37 Wood shingle coverage.

WOOD SHINGLES

NAME	LENGTH, THICKNESS, AND REFERENCE NOMENCLATURE	DESCRIPTION
NO.1** TAPER-SAWN	16" x .40" FIVEX (406mm x 10mm) 18" x .45" PERFECTIONS (457mm x 11mm) 24" x .50" ROYALS (610mm x 13mm)	TOP GRADE (NO. 1***) WOOD SHINGLES FOR USE AS ROOFING. THE WOOD BLANKS ARE RUN DIAGONALLY THROUGH A BANDSAW TO PRODUCE SHINGLES THAT ARE SAWN BOTH SIDES.
NO.1** FANCY SAWN BUTT	16" x .40" (406mm x 10mm) 18" x .45" (457mm x 11mm) 24" x .50" (610mm x 13mm)	TOP GRADE (NO. 1***) WOOD SHINGLE FOR USE AS ROOFING. TAPER-SAWN BOTH SIDES, WITH BUTT END CUT TO SPECIFIC SHAPE. A VARIETY OF FANCY BUTTS ARE AVAILABLE.

OCTAGON

HALF COVE

ROUND

DIAGONAL

ARROW

HEXAGONAL

FISH-SCALE

SQUARE

DIAMOND

** LOWER GRADES (e.g., NO. 2 AND NO. 3) ARE AVAILABLE AND ARE USED FOR STARTER OR UNDERCOURSING.

FIGURE A2-38 Wood shingle shapes and sizes.

SHAKE COVERAGE

NO. 1 GRADE SHAKE TYPE, LENGTH, AND THICKNESS IN INCHES (mm)		APPROXIMATE COVERAGE IN SQ. FT. (m²) FOR 5 BUNDLES WHEN SHAKES ARE APPLIED WITH AN AVERAGE 1/2" (13mm) SPACING AT THE FOLLOWING WEATHER EXPOSURES, IN INCHES (mm)				
		5" (127)	5 1/2" (140)	7 1/2" (191)	8 1/2" (216)	10" (254)
SHAKES	HEAVIES 24" X 3/4" (610 x 19)	—	—	75(b) (6.96)	85 (7.90)	100(c) (9.29)
	MEDIUMS 24" X 1/2" (610 x13)	—	—	75(b) (6.96)	85 (7.90)	100(c) (9.29)
NO.1 HANDSPLIT & RESAWN	HEAVIES 18" X 3/4" (457 x 19)	—	55(b) (5.11)	75(c) (6.96)	—	—
	MEDIUMS 18" X 1/2" (457 x 13)	—	55(b) (5.11)	75(c) (6.96)	—	—
NO.1 TAPER-SAWN	24" X 5/8" (610 x 16)	—	—	75(b) (6.96)	85 (7.90)	100(c) (9.29)
	18" X 5/8" (457 x 16)	—	55(b) (5.11)	75(c) (6.96)	—	—

FIGURE A2-39   Shake coverage and exposure table. (continued)

NO.1 TAPER-SPLIT 24" X 1/2" (610 x 13)	—	—	75(b) (6.96)	85 (7.90)	100(c) (9.29)
NO.1 STRAIGHT-SPLIT 18" X 3/8" (457 x 10)	—	65(b) (6.04)	90(c) (8.36)	—	—
HANDSPLIT STARTER 24" X 3/8" (610 x 10)	50 (4.65)	—	75(b) (6.96)	—	—
15" STARTER COURSE 15" X 3/8" (381 x 10)	USE WITH SHAKES APPLIED NOT OVER 7 1/2" (191mm) WEATHER EXPOSURE				
NO.2 24" X 3/8" (610 x 10)	USE WITH SHAKES APPLIED NOT OVER 10" (254mm) WEATHER EXPOSURE				

(a)   5 BUNDLES MAY COVER 100 SQ. FT. (9.29m) WHEN USED AS STARTER COURSE AT 10" (254mm) WEATHER EXPOSURE; 7 BUNDLES MAY COVER 100 SQ. FT. (9.29m) WHEN USED AS STARTER COURSE AT 7 1/2" (191mm) WEATHER EXPOSURE.
(b)   MAXIMUM RECOMMENDED WEATHER EXPOSURE FOR TRIPLE COVERAGE ROOF CONSTRUCTION.
(c)   MAXIMUM RECOMMENDED WEATHER EXPOSURE FOR DOUBLE COVERAGE ROOF CONSTRUCTION.
(d)   MAXIMUM RECOMMENDED WEATHER EXPOSURE.

NOTE - ALL DIMENSIONS ARE APPROXIMATE

FIGURE A2-39   Shake coverage and exposure table.

Roofing type	Minimum slope	Life years	Relative cost	Weight (pounds/100 sq ft)
Asphalt shingle	4	15–20	Low	200–300
Slate	5	100	High	750–4000
Wood shake	3	50	High	300
Wood shingle	3	25	Medium	150

FIGURE A2-40   Roofing materials.

Material	Expected life span
Asphalt shingles	15 to 30 years
Fiberglass shingles	20 to 30 years
Wood shingles	20 years
Wood shakes	50 years
Slate	Indefinite
Clay tiles	Indefinite
Copper	In excess of 35 years
Aluminum	35 years
Built-up roofing	5 to 20 years

Note: All estimated life spans depend on installation procedure, maintenance, and climatic conditions.

FIGURE A2-41   Potential life spans for various types of roofing materials.

Type of roofing	Weight in pounds
Clay shingle tile	1000–2000
Clay Spanish tile	800–1500
Slate	600–1600
Asphalt shingles	130–325
Wood shingles	200–300

FIGURE A2-42   Approximate weights of roofing materials, based on 100 square feet of material intalled.

Material	Slope
Asphalt or fiberglass shingle	4 in 12 slope
Roll roofing with exposed nails	3 in 12 slope
Roll roofing with concealed nails 3" head lap	2 in 12 slope
Double coverage half lap	1 in 12 slope

Lower slope: Treat as a flat roof. Use a continuous membrane system: either built up felt/asphalt with crushed stone or metal system with sealed or soldered seams.

FIGURE A2-43   Roofing materials and their lowest permissible slope.

Type	Use
Softwood veneer	Cross laminated plies or veneers—Sheathing, general construction and industrial use, etc.
Hardwood veneer	Cross laminated plies with hardwood face and back veneer—Furniture and cabinet work, etc.
Lumbercore plywood	Two face veneers and 2 crossband plies with an inner core of lumber strips—Desk and table tops, etc.
Medium-density overlay (MDO)	Exterior plywood with resin and fiber veneer—Signs, soffits, etc.
High-density overlay (HDO)	Tougher than MDO—Concrete forms, workbench tops, etc.
Plywood siding	T-111 and other textures used as one step sheathing and siding where codes allow.

FIGURE A2-44   Types and uses of plywood.

Material	Care	Life, yr	Cost
Aluminum	None	30	Medium
Hardboard	Paint Stain	30	Low
Horizontal wood	Paint Stain None	50+	Medium to high
Plywood	Paint Stain	20	Low
Shingles	Stain None	50+	High
Stucco	None	50+	Low to medium
Vertical wood	Paint Stain None	50+	Medium
Vinyl	None	30	Low

FIGURE A2-45 Siding material comparison.

Penny size "d"	Length	Approximate number per pound common	Approximate number per pound box	Approximate number per pound finish
2	1"	875	1000	1300
3	1¼"	575	650	850
4	1½"	315	450	600
5	1¾"	265	400	500
6	2"	190	225	300
7	2¼"	160		
8	2½"	105	140	200
9	2¾"	90		
10	3"	70	90	120
12	3¼"	60	85	110
16	3½"	50	70	90
20	4"	30	50	60
30	4½"	25		
40	5"	20		
50	5½"	15		
60	6"	10		

Note: Aluminum and c. c. nails are slightly smaller than other nails of the same penny size.

FIGURE A2-46   Nail sizes and number per pound.

Length	Gauge numbers
¼"	0 to 3
⅜"	2 to 7
½"	2 to 8
⅝"	3 to 10
¾"	4 to 11
⅞"	6 to 12
1"	6 to 14
1¼"	7 to 16
1½"	6 yo 18
1¾"	8 to 20
2"	8 to 20
2¼"	9 to 20
2½"	12 to 20
2¾"	14 to 20
3"	16 to 20
3½"	18 to 20
4"	18 to 20

FIGURE A2-47 Screw lengths and available gauge numbers.

Material	Length of material (in.)	Maximum exposure (in.)		
			Double coursing	
		Single coursing	No. 1 grade	No. 2 grade
Shingles	16	7-1/2	12	10
	18	8-1/2	14	11
	24	11-1/2	16	14
Shakes	18	8-1/2	14	–
(handsplit	24	11-1/2	20	–
and resawn)	32	15	–	–

FIGURE A2-48   Exposure distances for wood shingles and shakes on side walls.

Height	Width
80"	36"
84"	36"
80"	34"
84"	34"
80"	32"
84"	32"

FIGURE A2-49   Stock sizes of exterior doors.

Wind velocity (mph)	Window glass thickness (in.) @ 0.133
30	64.5 square feet
40	32.25 square feet
65	16.1 square feet
120	4.5 square feet

FIGURE A2-50   Maximum range of glass size, based on wind velocity.

Wind velocity (mph)	Window glass thickness (in.) @ 0.085
30	30 square feet
40	17.5 square feet
65	8.5 square feet
120	2.5 square feet

FIGURE A2-51   Maximum range of glass size, based on wind velocity.

Heat Gain and Performance Data

Heat Gain Data

In areas of the U.S. where cooling is the major energy cost, glazing may be the most important factor in energy-saving. That's because cooling costs are based almost solely on heat gains transmitted through the glass. The accompanying table is used to show maximum heat gain by type of glass.

Clear	Heat Gain	Tinted Grey/Bronze	Heat Gain	Medium Performance Reflective	Heat Gain
Single-pane ½" or ¼"	214	Single-pane grey ¾" (for comparison only)	165	Single-pane bronze (for comparison only)	106
Single-pane ¾" (for comparison only)	208	Single pane bronze ¾" (for comparison only)	168		
Double-pane (for comparison only)	186				
Double-pane high-performance insulating	113	Double-pane high-1 performance sun insulating			

FIGURE A2-52   Heat gain for various types of glazing.

STANDARD OPENING SIZE

Opening Widths	Opening Heights							1 3/8 " Doors	
	1 3/4 " Doors								
2'0"	6'8"	7'0"	7'2"	7'10"	8'0"	8'10"	10'0"	6'8"	7'0"
2'4"	6'8"	7'0"	7'2"	7'10"	8'0"	8'10"	10'0"	6'8"	7'0"
2'6"	6'8"	7'0"	7'2"	7'10"	8'0"	8'10"	10'0"	6'8"	7'0"
2'8"	6'8"	7'0"	7'2"	7'10"	8'0"	8'10"	10'0"	6'8"	7'0"
2'10"	6'8"	7'0"	7'2"	7'10"	8'0"	8'10"	10'0"	6'8"	7'0"
3'0"	6'8"	7'0"	7'2"	7'10"	8'0"	8'10"	10'0"	6'8"	7'0"
3'4"	6'8"	7'0"	7'2"	7'10"	8'0"	8'10"	10'0"		
3'6"	6'8"	7'0"	7'2"	7'10"	8'0"	8'10"	10'0"		
3'8"	6'8"	7'0"	7'2"	7'10"	8'0"	8'10"	10'0"		
3'10"	6'8"	7'0"	7'2"	7'10"	8'0"	8'10"	10'0"		
4'0"	6'8"	7'0"	7'2"	7'10"	8'0"	8'10"	10'0"		

FIGURE A2-53   Standard opening size for hollow metal doors.

Glass size (in inches)	Frame size (width × height, feet and inches)	Rough opening (width × height, feet and inches)
33 × 76¾	6-0 × 6-10¾	6-0½ × 6-11¼
45 × 76¾	8-0 × 6-10¾	8-0½ × 6-11¼
57 × 76¾	10-0 × 6-10¾	10-0½ × 6-11¼
33 × 76¾	9-0 × 6-10¾	8-0½ × 6-11¼
45 × 76¾	12-0 × 6-10¾	12-0½ × 6-11¼
57 × 76¾	15-0 × 6-10¾	15-0½ × 6-11¼
33 × 76¾	11-11 × 6-10¾	11-11½ × 6-11¼
45 × 76¾	15-11 × 6-10¾	15-11½ × 6-11¼
57 × 76¾	19-11 × 6-10¾	19-11½ × 6-11¼

FIGURE A2-54   Measurements for sliding-glass doors.

Passageway	Recommended	Minimum
Stairs	40"	36"
Landings	40"	36"
Main hall	48"	36"
Minor hall	36"	30"
Interior door	32"	28"
Exterior door	36"	36"

FIGURE A2-55   Widths of passageways.

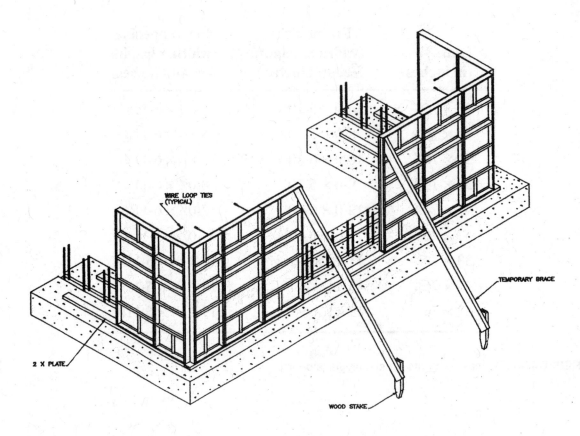

FIGURE A2-56 Typical concrete wall from schematic with one side in place.

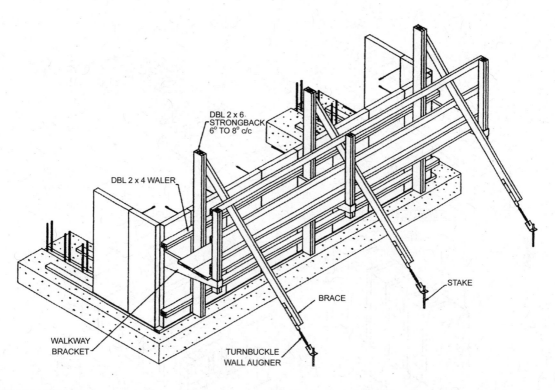

FIGURE A2-57   Typical concrete wall from schematic with walkway bracket installed and one side in place.

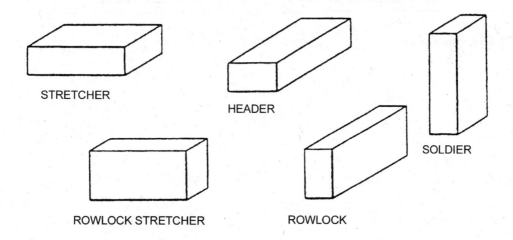

FIGURE A2-58   Brick positions in a wall.

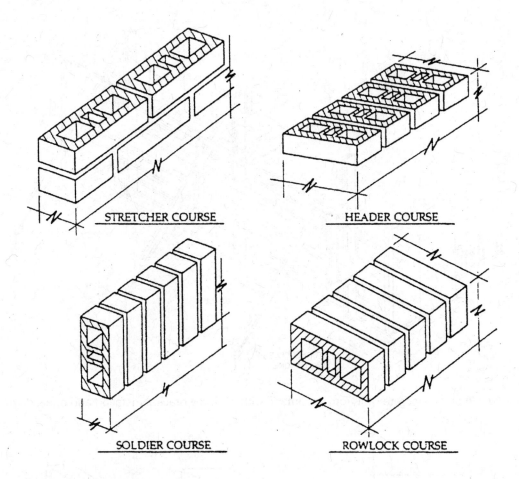

STRETCHER COURSE

HEADER COURSE

SOLDIER COURSE

ROWLOCK COURSE

FIGURE A2-59 Brick orientation.

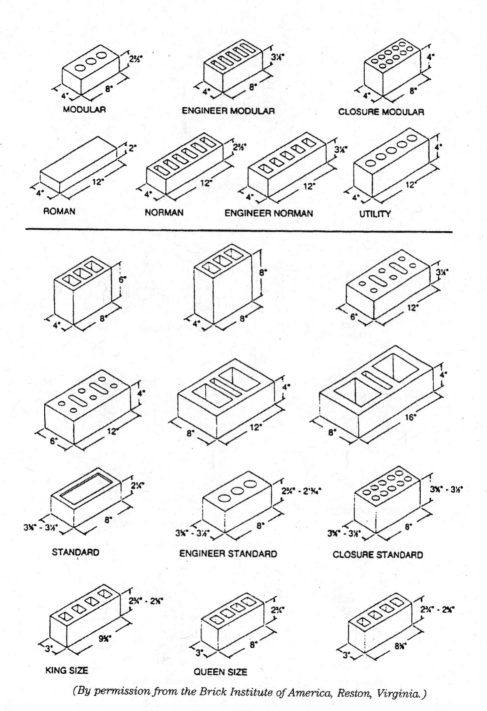

MODULAR

ENGINEER MODULAR

CLOSURE MODULAR

ROMAN

NORMAN

ENGINEER NORMAN

UTILITY

STANDARD

ENGINEER STANDARD

CLOSURE STANDARD

KING SIZE

QUEEN SIZE

(By permission from the Brick Institute of America, Reston, Virginia.)

FIGURE A2-60   Modular (above) and nonmodular (below) brick sizes.

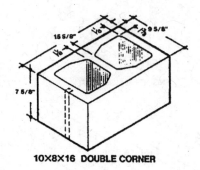

10X8X16 DOUBLE CORNER

Miscellaneous Shapes

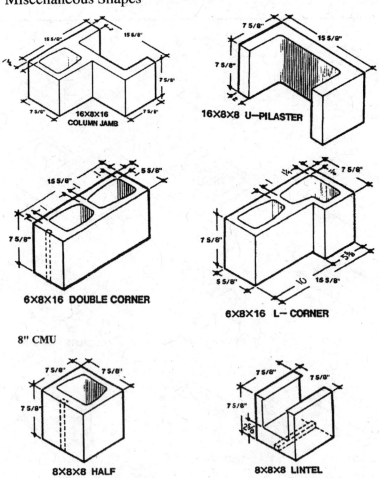

16X8X16 COLUMN JAMB

16X8X8 U—PILASTER

6X8X16 DOUBLE CORNER

6X8X16 L—CORNER

8" CMU

8X8X8 HALF

8X8X8 LINTEL

FIGURE A2-61 CMU shapes and sizes.

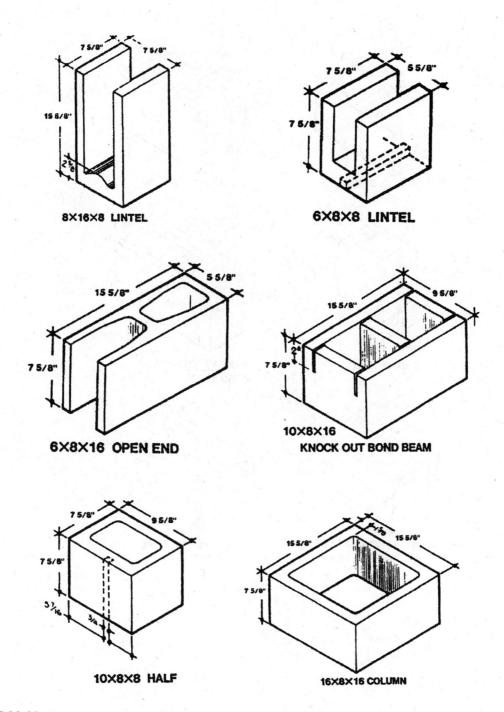

8X16X8 LINTEL

6X8X8 LINTEL

6X8X16 OPEN END

10X8X16
KNOCK OUT BOND BEAM

10X8X8 HALF

16X8X16 COLUMN

FIGURE A2-62 CMU shapes and sizes.

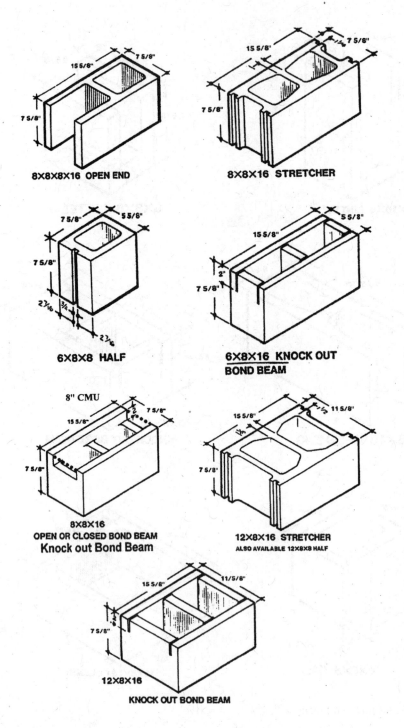

FIGURE A2-63   CMU shapes and sizes.

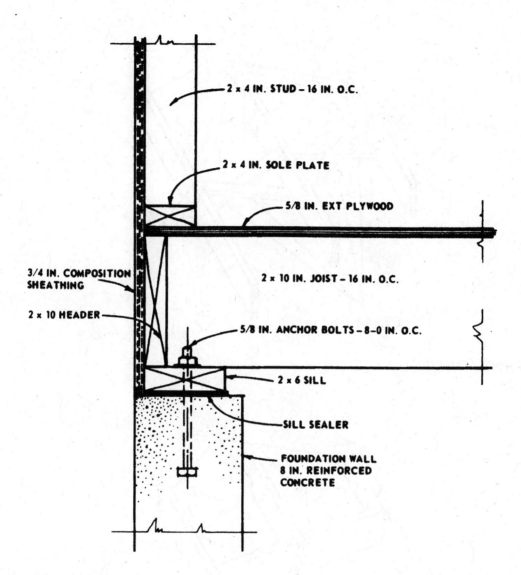

FIGURE A2-64   Framing detail.

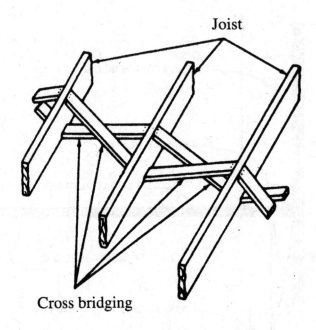

Joist

Cross bridging

FIGURE A2-65 Cross bridging.

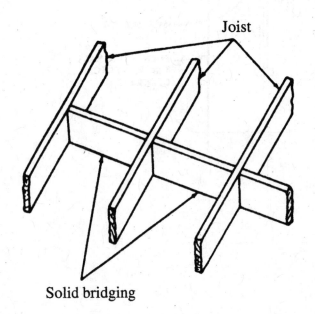

Joist

Solid bridging

FIGURE A2-66 Solid bridging.

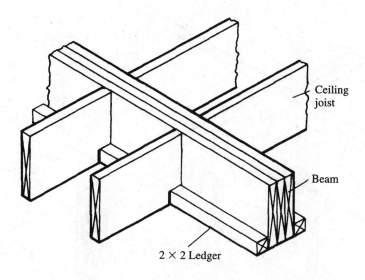

FIGURE A2-67   The use of a ledger support with a wooden beam (courtesy of U.S. government).

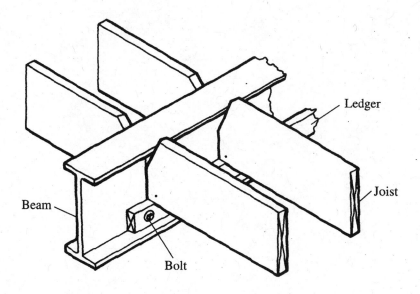

FIGURE A2-68   The use of a ledger support with a steel beam (courtesy of U.S. government).

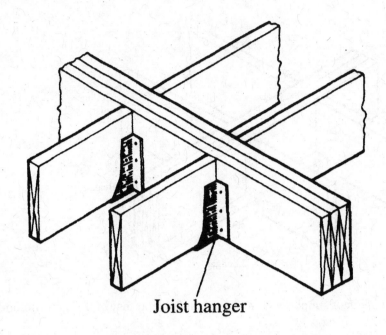

Joist hanger

FIGURE A2-69   Joist hangers in use (courtesy of U.S. government).

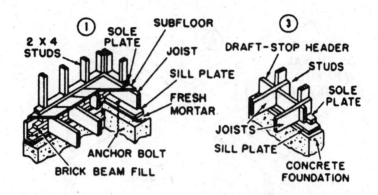

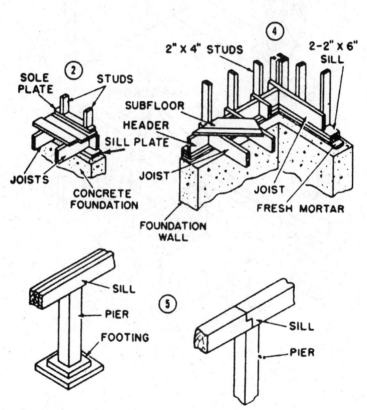

FIGURE A2-70 Types of sills.

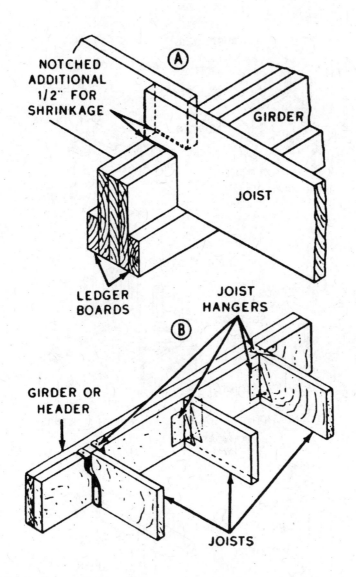

FIGURE A2-71 Joist to girder attachment.

Without sill plate

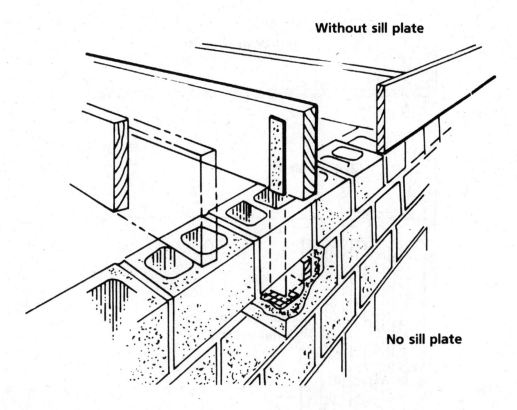

No sill plate

FIGURE A2-72 Floor joist detail without a sill plate (courtesy of USDA Forest Service).

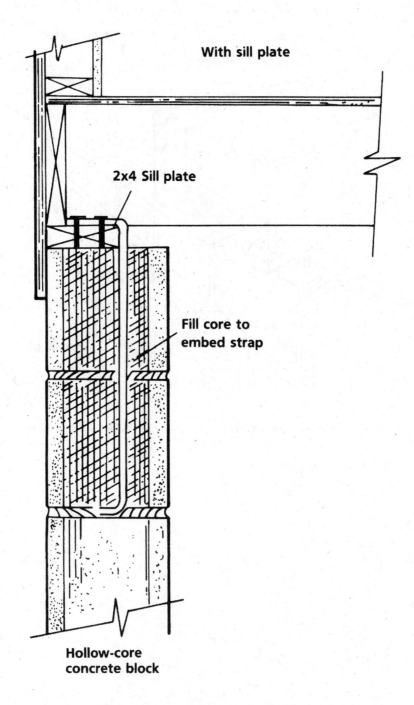

With sill plate

2x4 Sill plate

Fill core to embed strap

Hollow-core concrete block

FIGURE A2-73 Floor joist detail with a sill plate (courtesy of USDA Forest Service).

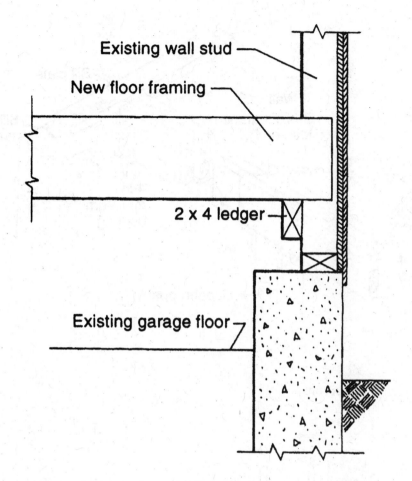

FIGURE A2-74   Framing to bring a floor in an addition, existing porch, or garage up to the same floor level as a home.

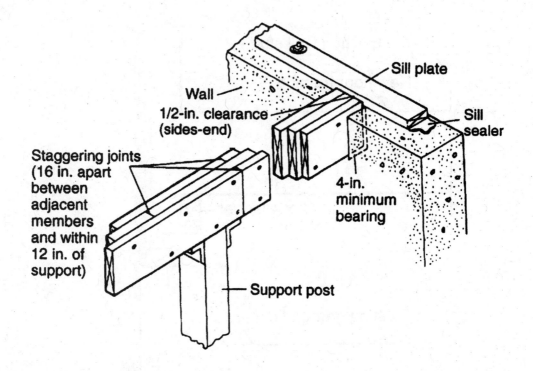

FIGURE A2-75 Beam construction.

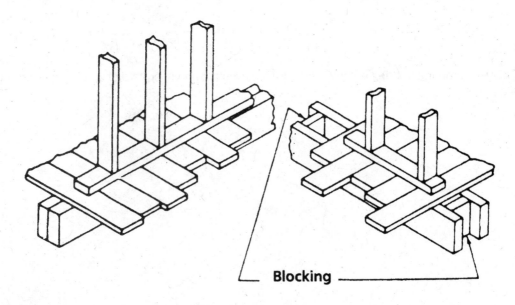

FIGURE A2-76 Reinforced joist installation (courtesy of USDA Forest Service).

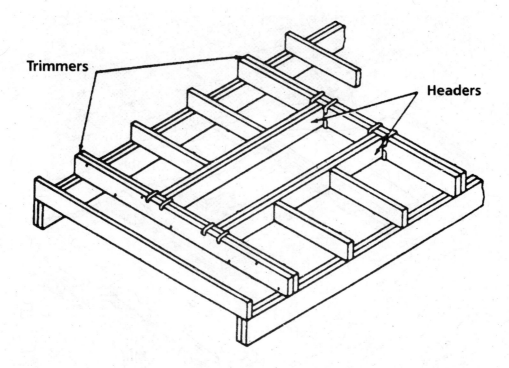

FIGURE A2-77   The use of headers to reinforce joists where a hole is needed, such as for a stairway (courtesy of USDA Forest Service).

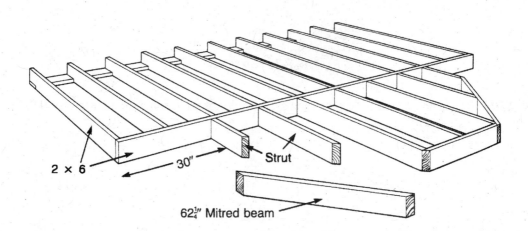

FIGURE A2-78   Sleeper system used to build over existing concrete (courtesy of Georgia-Pacific Corp.).

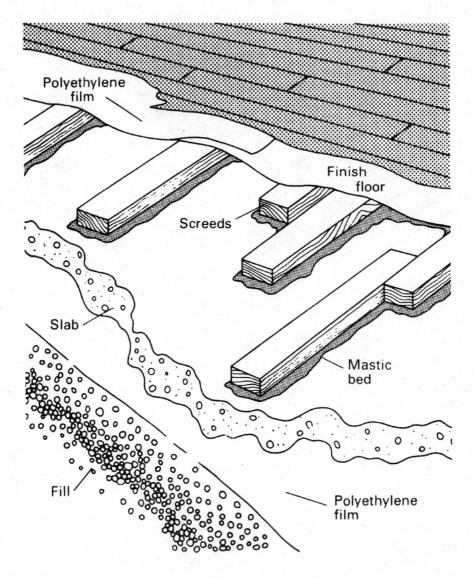

FIGURE A2-79   Using screeds to build a flooring system over existing concrete (courtesy of U.S. Dept. of Agriculture).

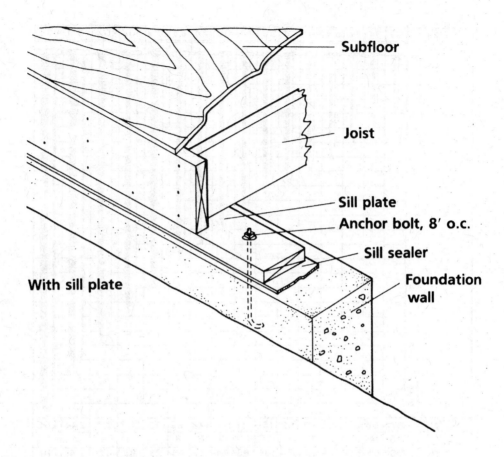

Subfloor

Joist

Sill plate

Anchor bolt, 8' o.c.

Sill sealer

Foundation wall

With sill plate

FIGURE A2-80   Typical floor structure with a plywood panel (courtesy of USDA Forest Service).

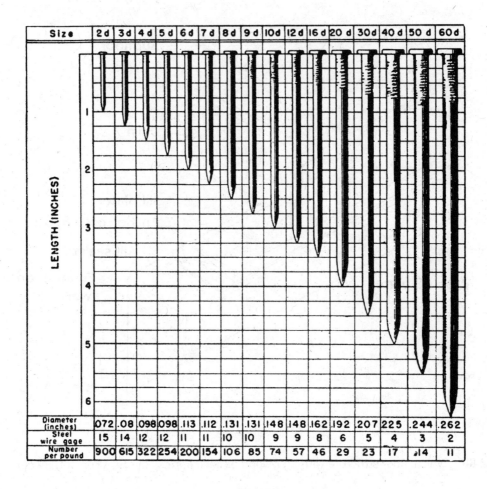

Size	2d	3d	4d	5d	6d	7d	8d	9d	10d	12d	16d	20d	30d	40d	50d	60d
Diameter (inches)	.072	.08	.098	.098	.113	.112	.131	.131	.148	.148	.162	.192	.207	.225	.244	.262
Steel wire gage	15	14	12	12	11	11	10	10	9	9	8	6	5	4	3	2
Number per pound	900	615	322	254	200	154	106	85	74	57	46	29	23	17	14	11

FIGURE A2-81   Nail sizes (courtesy of USDA Forest Service).

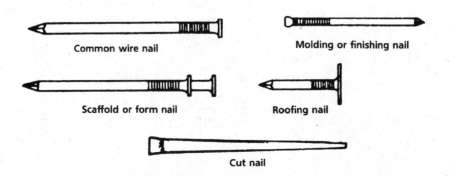

Common wire nail

Molding or finishing nail

Scaffold or form nail

Roofing nail

Cut nail

FIGURE A2-82   Types of nails (courtesy of USDA Forest Service).

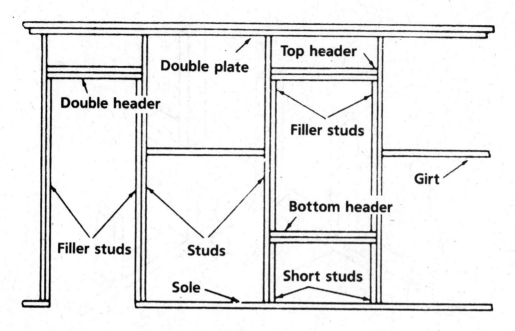

FIGURE A2-83 Typical door and window framing that requires additoinal material during the framing process (courtesy of USDA Forest Service).

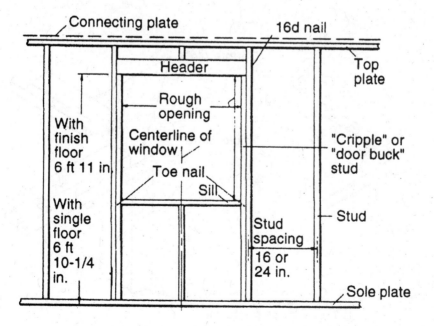

FIGURE A2-84 Typical framing with a header over the rough opening for a window.

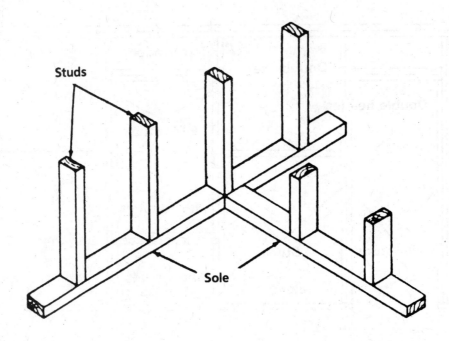

FIGURE A2-85   The sole plate is attached to the floor system (courtesy of USDA Forest Service).

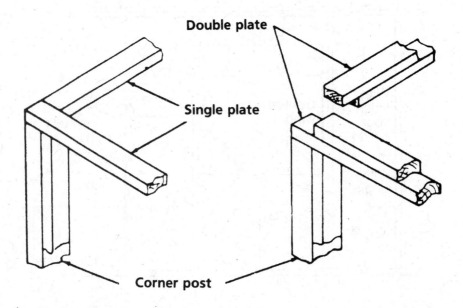

FIGURE A2-86   Examples of a single top plate and a double top plate (courtesy of USDA Forest Service).

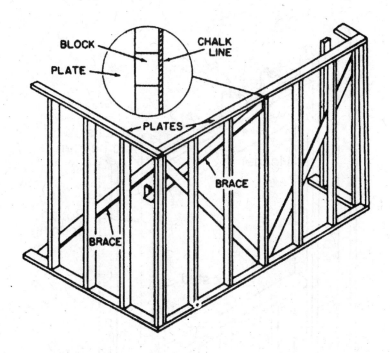

FIGURE A2-87 Wall braces for exterior stud walls.

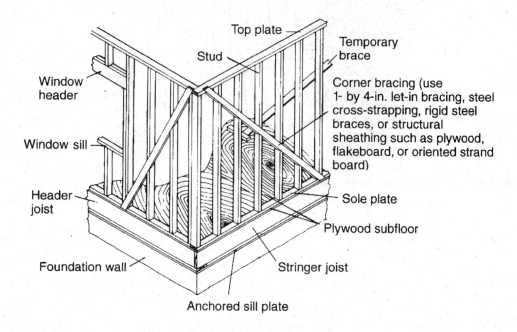

FIGURE A2-88 Wall framing with platform construction.

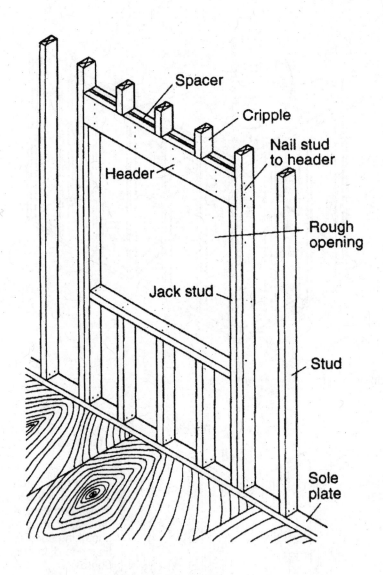

FIGURE A2-89   Typical rough opening for a window.

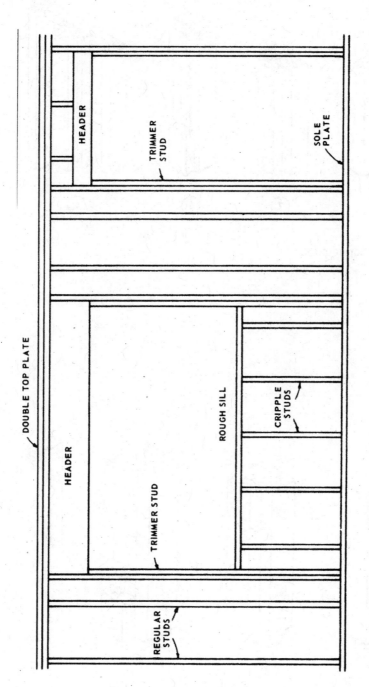

FIGURE A2-90 Typical rough opening for a door and window.

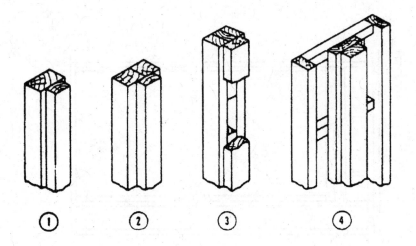

FIGURE A2-91   T-post connectors.

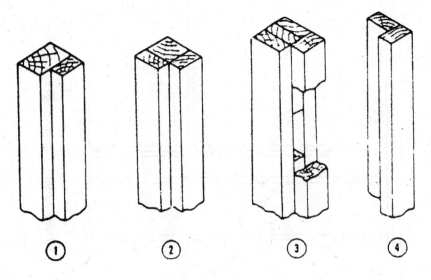

FIGURE A2-92   Corner posts.

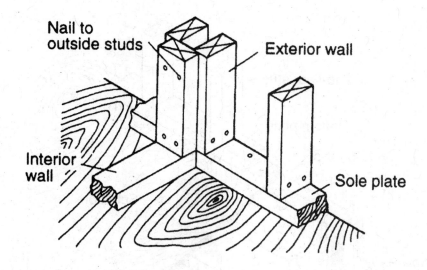

FIGURE A2-93 Tying a partition wall to an exterior wall.

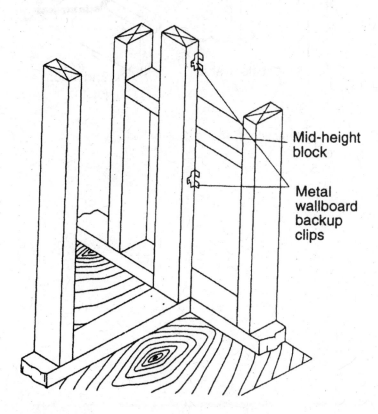

FIGURE A2-94 Using blocking as a means for attaching wall sections.

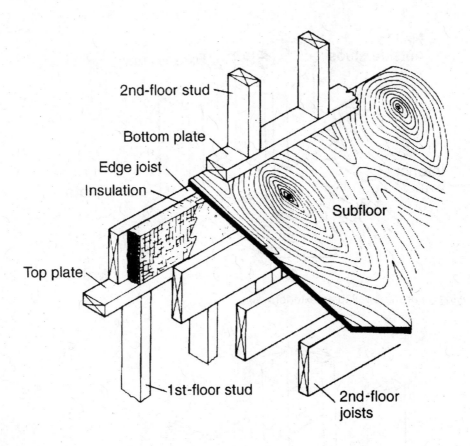

FIGURE A2-95   Second-story framing for platform construction.

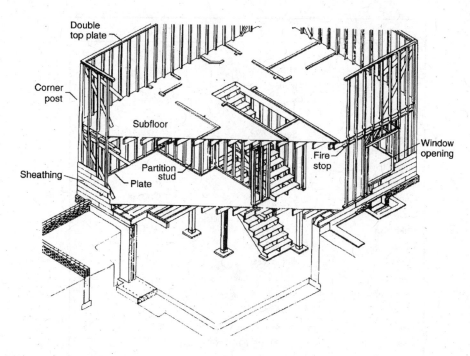

FIGURE A2-96   Wall framing for a two-story home.

To change	To	Multiply by
Inches	Feet	0.0833
Inches	Millimeters	25.4
Feet	Inches	12
Feet	Yards	0.3333
Yards	Feet	3
Square inches	Square feet	0.00694
Square feet	Square inches	144
Square feet	Square yards	0.11111
Square yards	Square feet	9
Cubic inches	Cubic feet	0.00058
Cubic feet	Cubic inches	1728
Cubic feet	Cubic yards	0.03703
Gallons	Cubic inches	231
Gallons	Cubic feet	0.1337
Gallons	Pounds of water	8.33
Pounds of water	Gallons	0.12004
Ounces	Pounds	0.0625
Pounds	Ounces	16
Inches of water	Pounds per square inch	0.0361
Inches of water	Inches of mercury	0.0735
Inches of water	Ounces per square inch	0.578
Inches of water	Pounds per square foot	5.2
Inches of mercury	Inches of water	13.6
Inches of mercury	Feet of water	1.1333
Inches of mercury	Pounds per square inch	0.4914
Ounces per square inch	Inches of mercury	0.127
Ounces per square inch	Inches of water	1.733
Pounds per square inch	Inches of water	27.72
Pounds per square inch	Feet of water	2.310
Pounds per square inch	Inches of mercury	2.04
Pounds per square inch	Atmospheres	0.0681
Feet of water	Pounds per square inch	0.434
Feet of water	Pounds per square foot	62.5
Feet of water	Inches of mercury	0.8824
Atmospheres	Pounds per square inch	14.696
Atmospheres	Inches of mercury	29.92
Atmospheres	Feet of water	34
Long tons	Pounds	2240
Short tons	Pounds	2000
Short tons	Long tons	0.89295

FIGURE A2-97 Measurement conversion factors.

To find	Multiply	By
Microns	Mils	25.4
Centimeters	Inches	2.54
Meters	Feet	0.3048
Meters	Yards	0.19144
Kilometers	Miles	1.609344
Grams	Ounces	28.349523
Kilograms	Pounds	0.4539237
Liters	Gallons (U.S.)	3.7854118
Liters	Gallons (Imperial)	4.546090
Milliliters (cc)	Fluid ounces	29.573530
Milliliters (cc)	Cubic inches	16.387064
Square centimeters	Square inches	6.4516
Square meters	Square feet	0.09290304
Square meters	Square yards	0.83612736
Cubic meters	Cubic feet	2.8316847×10^{-2}
Cubic meters	Cubic yards	0.76455486
Joules	BTU	1054.3504
Joules	Foot-pounds	1.35582
Kilowatts	BTU per minute	0.01757251
Kilowatts	Foot-pounds per minute	2.2597×10^{-5}
Kilowatts	Horsepower	0.7457
Radians	Degrees	0.017453293
Watts	BTU per minute	17.5725

FIGURE A2-98 Conversion factors in converting from customary (U.S.) units to metric units.

	Imperial	Metric
Length	1 inch	25.4 mm
	1 foot	0.3048 m
	1 yard	0.9144 m
	1 mile	1.609 km
Mass	1 pound	0.454 kg
	1 U.S. short ton	0.9072 tonne
Area	1 ft^2	0.092 m^2
	1 yd^2	0.836 m^2
	1 acre	0.404 hectare (ha)
Capacity/Volume	1 ft^3	0.028 m^3
	1 yd^3	0.764 m^3
	1 liquid quart	0.946 litre (1)
	1 gallon	3.785 litre (1)
Heat	1 BTU	1055 joule (J)
	1 BTU/hr	0.293 watt (W)

FIGURE A2-99   Measurement conversions: Imperial to metric.

Area = Short diameter × long diameter × .7854

FIGURE A2-100   Ellipse calcuation.

Area of surface = One half of circumference of base × slant height + area
of base.
Volume = Diameter × diameter × .7854 × one-third of the altitude.

FIGURE A2-101   Cone calculation.

Volume = Width × height × length

FIGURE A2-102   Volume of a rectangular prism.

Area = Length × width

FIGURE A2-103   Finding the area of a square.

Area = ½ perimeter of base × slant height + area of base
Volume = Area of base × ⅓ of the altitude

FIGURE A2-104   Finding area and volume of a pyramid.

These comprise the numerous figures having more than four sides, names according to the number of sides, thus:

Figure	Sides
Pentagon	5
Hexagon	6
Heptagon	7
Octagon	8
Nonagon	9
Decagon	10

To find the area of a polygon: Multiply the sum of the sides (perimeter of the polygon) by the perpendicular dropped from its center to one of its sides, and half the product will be the area. This rule applies to all regular polygons.

FIGURE A2-105   Polygons.

$$\text{Area} = \text{Width of side} \times 2.598 \times \text{width of side}$$

FIGURE A2-106   Hexagons.

$$\text{Area} = \text{Base} \times \text{distance between the two parallel sides}$$

FIGURE A2-107   Parallelograms.

Area of surface = Diameter × diameter × 3.1416
Side of inscribed cube = Radius × 1.547
Volume = Diameter × diameter × diameter × .5236

FIGURE A2-108   Spheres.

Area = One-half of height times base

FIGURE A2-109   Triangles.

Area = One-half of the sum of the parallel sides × the height

FIGURE A2-110   Trapezoids.

Volume = Width × height × length

FIGURE A2-111   Cubes.

Circumference = Diameter × 3.1416
Circumference = Radius × 6.2832
Diameter = Radius × 2
Diameter = Square root of (area ÷ .7854)
Diameter = Square root of area × 1.1283
Diameter = Circumference × .31831
Radius = Diameter ÷ 2
Radius = Circumference × .15915
Radius = Square root of area × .56419
Area = Diameter × Diameter × .7854
Area = Half of the circumference × half of the diameter
Area = Square of the circumference × .0796
Arc length = Degrees × radius × .01745
Degrees of arc = Length ÷ (radius × .01745)
Radius of arc = Length ÷ (degrees × .01745)
Side of equal square = Diameter × .8862
Side of inscribed square = Diameter × .7071
Area of sector = Area of circle × degrees of arc ÷ 360

FIGURE A2-112   Formulas for a circle.

1. Circumference of a circle = $\pi \times$ diameter or 3.1416 $\times$ diameter
2. Diameter of a circle = Circumference $\times$ .31831
3. Area of a square = Length $\times$ width
4. Area of a rectangle = Length $\times$ width
5. Area of a parallelogram = Base $\times$ perpendicular height
6. Area of a triangle = ½ base $\times$ perpendicular height
7. Area of a circle = $\pi \times$ radius squared or diameter squared $\times$ .7854
8. Area of an ellipse = Length $\times$ width $\times$ .7854
9. Volume of a cube or rectangular prism = Length $\times$ width $\times$ height
10. Volue of a triangular prism = Area of triangle $\times$ length
11. Volume of a sphere = Diameter cubed $\times$ .5236 or (dia. $\times$ dia. $\times$ dia. $\times$ .5236)
12. Volume of a cone = $\pi \times$ radius square $\times$ ⅓ height
13. Volume of a cylinder = $\pi \times$ radius squared $\times$ height
14. Length of one side of a square $\times$ 1.128 = Diameter of an equal circle
15. Doubling the diameter of a pipe or cylinder increases its capacity 4 times
16. The pressure (in lbs. per sq. inch) of a column of water = Height of the column (in feet) $\times$ .434
17. The capacity of a pipe or tank (in U.S. gallons) = Diameter squared (in inches) $\times$ the length (in inches) $\times$ .0034
18. A gallon of water = 8⅓ lb. = 231 cu. inches
19. A cubic foot of water = 62½ lb. = 7½ gallons

FIGURE A2-113 Useful formulas.

MULTIPLY	BY	TO OBTAIN
Gallons/minute	8.0208	Cubic feet/hour
Gallons water/minute	6.0086	Tons of water/24 hours
Inches	2.540	Centimeters
Inches of mercury	0.03342	Atmospheres
Inches of mercury	1.133	Feet of water
Inches of mercury	0.4912	Pounds/square inch
Inches of water	0.002458	Atmospheres
Inches of water	0.07355	Inches of mercury
Inches of water	5.202	Pounds/square feet
Inches of water	0.03613	Pounds/square inch
Liters	1000	Cubic centimeters
Liters	61.02	Cubic inches
Liters	0.2642	Gallons
Miles	5280	Feet
Miles/hour	88	Feet/minute
Miles/hour	1.467	Feet/second
Millimeters	0.1	Centimeters
Millimeters	0.03937	Inches
Million gallon/day	1.54723	Cubic feet/second
Pounds of water	0.01602	Cubic feet
Pounds of water	27.68	Cubic inches
Pounds of water	0.1198	Gallons
Pounds/cubic inch	1728	Pounds/cubic feet
Pounds/square foot	0.01602	Feet of water
Pounds/square inch	0.06804	Atmospheres
Pounds/square inch	2.307	Feet of water
Pounds/square inch	2.036	Inches of mercury
Quarts (dry)	67.20	Cubic inches
Quarts (liquid)	57.75	Cubic inches
Square feet	144	Square inches
Square miles	640	Acres
Square yards	9	Square feet
Temperature (°C) + 273	1	Abs. temperature (°C)
Temperature (°C) + 17.28	1.8	Temperature (°F)
Temperature (°F) + 460	1	Abs. temperature (°F)
Temperature (°F) - 32	5/9	Temperature (°C)
Tons (short)	2000	Pounds
Tons of water/24 hours	83.333	Pounds water/hour
Tons of water/24 hours	0.16643	Gallons/minute
Tons of water/24 hours	1.3349	Cubic feet/hour

FIGURE A2-114   A useful set of tables to keep on hand (reprinted from the 2000 Uniform Plumbing Code [UPC] with the permission of the International Association of Plumbing and Mechanical Officials [IAPMO]).

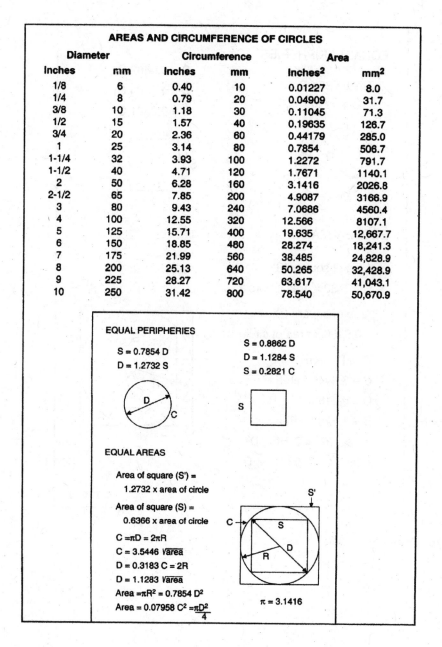

AREAS AND CIRCUMFERENCE OF CIRCLES

Diameter		Circumference		Area	
Inches	mm	Inches	mm	Inches²	mm²
1/8	6	0.40	10	0.01227	8.0
1/4	8	0.79	20	0.04909	31.7
3/8	10	1.18	30	0.11045	71.3
1/2	15	1.57	40	0.19635	126.7
3/4	20	2.36	60	0.44179	285.0
1	25	3.14	80	0.7854	506.7
1-1/4	32	3.93	100	1.2272	791.7
1-1/2	40	4.71	120	1.7671	1140.1
2	50	6.28	160	3.1416	2026.8
2-1/2	65	7.85	200	4.9087	3166.9
3	80	9.43	240	7.0686	4560.4
4	100	12.55	320	12.566	8107.1
5	125	15.71	400	19.635	12,667.7
6	150	18.85	480	28.274	18,241.3
7	175	21.99	560	38.485	24,828.9
8	200	25.13	640	50.265	32,428.9
9	225	28.27	720	63.617	41,043.1
10	250	31.42	800	78.540	50,670.9

EQUAL PERIPHERIES

$S = 0.7854\ D$

$D = 1.2732\ S$

$S = 0.8862\ D$

$D = 1.1284\ S$

$S = 0.2821\ C$

EQUAL AREAS

Area of square (S') =

   1.2732 x area of circle

Area of square (S) =

   0.6366 x area of circle

$C = \pi D = 2\pi R$

$C = 3.5446\ \sqrt{area}$

$D = 0.3183\ C = 2R$

$D = 1.1283\ \sqrt{area}$

$Area = \pi R^2 = 0.7854\ D^2$

$Area = 0.07958\ C^2 = \dfrac{\pi D^2}{4}$

$\pi = 3.1416$

FIGURE A2-115 More useful information (reprinted from the 2000 Uniform Plumbing Code [UPC] with the permission of the International Association of Plumbing and Mechanical Officials [IAPMO]).

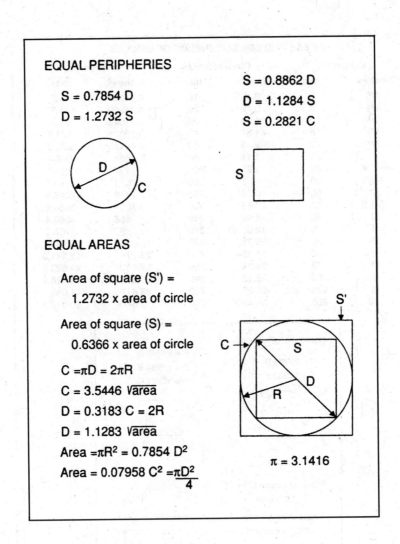

EQUAL PERIPHERIES

$S = 0.7854\ D$
$D = 1.2732\ S$

$S = 0.8862\ D$
$D = 1.1284\ S$
$S = 0.2821\ C$

EQUAL AREAS

Area of square (S') =
   1.2732 x area of circle

Area of square (S) =
   0.6366 x area of circle

$C = \pi D = 2\pi R$

$C = 3.5446\ \sqrt{area}$

$D = 0.3183\ C = 2R$

$D = 1.1283\ \sqrt{area}$

Area $= \pi R^2 = 0.7854\ D^2$

Area $= 0.07958\ C^2 = \dfrac{\pi D^2}{4}$

$\pi = 3.1416$

FIGURE A2-116 Mathematical formulas.

Parallelogram	Area = base × distance between the two parallel sides
Pyramid	Area = ½ perimeter of base × slant height + area of base
	Volume = area of base × ⅓ of the altitude
Rectangle	Area = length × width
Rectangular prism	Volume = width × height × length
Sphere	Area of surface = diameter × diameter × 3.1416
	Side of inscribed cube = radius × 1.547
	Volume = diameter × diameter × diameter × 0.5236
Square	Area = length × width
Triangle	Area = one-half of height times base
Trapezoid	Area = one-half of the sum of the parallel sides × the height
Cone	Area of surface = one-half of circumference of base × slant height + area of base
	Volume = diameter × diameter × 0.7854 × one-third of the altitude
Cube	Volume = width × height × length
Ellipse	Area = short diameter × long diameter × 0.7854
Cylinder	Area of surface = diameter × 3.1416 × length + area of the two bases
	Area of base = diameter × diameter × 0.7854
	Area of base = volume ÷ length
	Length = volume ÷ area of base
	Volume = length × area of base
	Capacity in gallons = volume in inches ÷ 231
	Capacity of gallons = diameter × diameter × length × 0.0034
	Capacity in gallons = volume in feet × 7.48
Circle	Circumference = diameter × 3.1416
	Circumference = radius × 6.2832
	Diameter = radius × 2
	Diameter = square root of = (area ÷ 0.7854)
	Diameter = square root of area × 1.1233

FIGURE A2-117   Area and other formulas.

−100°–30°		
°C	*Base temperature*	°F
−73	−100	−148
−68	−90	−130
−62	−80	−112
−57	−70	−94
−51	−60	−76
−46	−50	−58
−40	−40	−40
−34.4	−30	−22
−28.9	−20	−4
−23.3	−10	14
−17.8	0	32
−17.2	1	33.8
−16.7	2	35.6
−16.1	3	37.4
−15.6	4	39.2
−15.0	5	41.0
−14.4	6	42.8
−13.9	7	44.6
−13.3	8	46.4
−12.8	9	48.2
−12.2	10	50.0
−11.7	11	51.8
−11.1	12	53.6
−10.6	13	55.4
−10.0	14	57.2
31°–71°		
°C	*Base temperature*	°F
−0.6	31	87.8
0	32	89.6
0.6	33	91.4
1.1	34	93.2
1.7	35	95.0
2.2	36	96.8
2.8	37	98.6
3.3	38	100.4
3.9	39	102.2
4.4	40	104.0
5.0	41	105.8
5.6	42	107.6

FIGURE A2-118   Temperature conversion.

Vacuum in inches of mercury	Boiling point
29	76.62
28	99.93
27	114.22
26	124.77
25	133.22
24	140.31
23	146.45
22	151.87
21	156.75
20	161.19
19	165.24
18	169.00
17	172.51
16	175.80
15	178.91
14	181.82
13	184.61
12	187.21
11	189.75
10	192.19
9	194.50
8	196.73
7	198.87
6	200.96
5	202.25
4	204.85
3	206.70
2	208.50
1	210.25

FIGURE A2-119   Boiling points of water based on pressure.

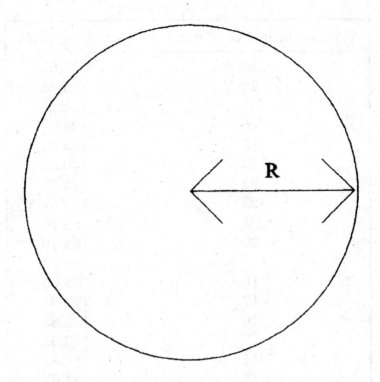

FIGURE A2-120   Radius of a circle.

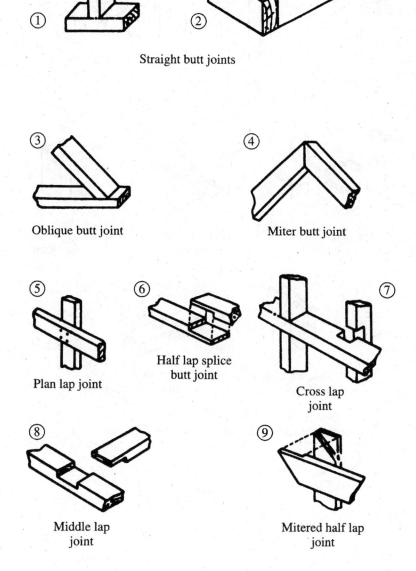

Straight butt joints

Oblique butt joint

Miter butt joint

Plan lap joint

Half lap splice
butt joint

Cross lap
joint

Middle lap
joint

Mitered half lap
joint

FIGURE A2-121 Butt and lap joints.

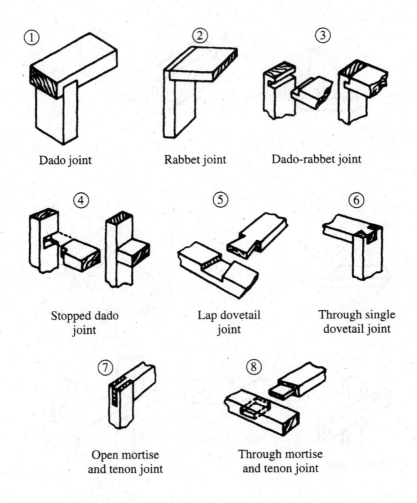

FIGURE A2-122   Dado, rabbet, dovetail, and mortise and tenon joints.

- Crown
- Rabbeted half round
- Half round
- Corner bead
- Sliding door
- Handrail
- Cove
- Quarter round
- Dowel
- Picture rail
- Scoop
- Edge

FIGURE A2-123   Types of trim molding.

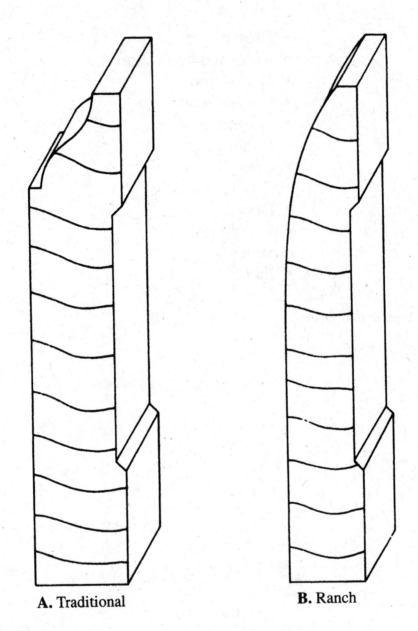

A. Traditional   **B.** Ranch

FIGURE A2-124   Base molding.

GENERAL MILLWORK PROFILES and NOMENCLATURE:

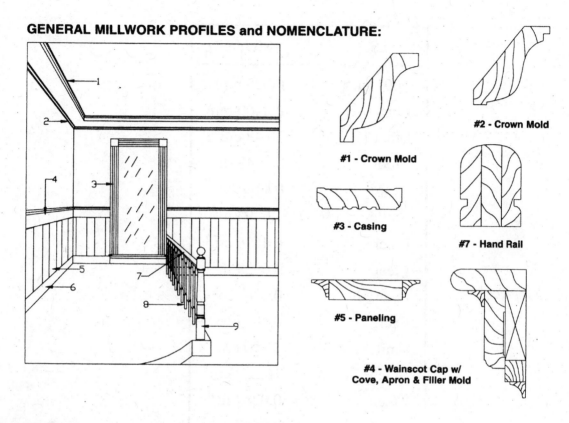

#1 - Crown Mold

#2 - Crown Mold

#3 - Casing

#7 - Hand Rail

#5 - Paneling

#4 - Wainscot Cap w/
Cove, Apron & Filler Mold

FIGURE A2-125 General millwork profiles and uses.

U.S.	Metric
0.001 inch	0.025 mm
1 inch	25.400 mm
1 foot	30.48 cm
1 foot	0.3048 m
1 yard	0.9144 m
1 mile	1.609 km
1 inch^2	6.4516 cm^2
1 feet^2	0.0929 m^2
1 yard^2	0.8361 m^2
1 acre	0.4047 ha
1 mile^2	2.590 km^2
1 inch^3	16.387 cm^3
1 feet^3	0.0283 m^3
1 yard^3	0.7647 m^3
1 U.S. ounce	29.57 ml
1 U.S. pint	0.4732 l
1 U.S. gallon	3.785 l
1 ounce	28.35 g
1 pound	0.4536 kg

FIGURE A2-126 Conversion tables.

Unit	Equals
1 meter	39.3 inches 3.28083 feet 1.0936 yards
1 centimeter	.3937 inch
1 millimeter	.03937 inch, or nearly ⅕ inch
1 kilometer	0.62137 mile
1 foot	.3048 meter
1 inch	2.54 centimeters 25.40 millimeters
1 square meter	10.764 square feet 1.196 square yards
1 square centimeter	.155 square inch
1 square millimeter	.00155 square inch
1 square yard	.836 square meter
1 square foot	.0929 square meter
1 square inch	6.452 square centimeter 645.2 square millimeter

FIGURE A2-127   Metric-customary equivalents.

Unit	Equals
1 cubic meter	35.314 cubic feet 1.308 cubic yards 264.2 U.S. gallons (231 cubic inches)
1 cubic decimeter	61.0230 cubic inches .0353 cubic feet
1 cubic centimeter	.061 cubuic inch
1 liter	1 cubic decimeter 61.0230 cubic inches 0.0353 cubic foot 1.0567 quarts (U.S.) 0.2642 gallon (U.S.) 2.2020 lb. of water at 62°F.
1 cubic yard	.7645 cubic meter
1 cubic foot	.02832 cubic meter 28.317 cubic decimeters 28.317 liters
1 cubic inch	16.383 cubic centimeters
1 gallon (British)	4.543 liters
1 gallon (U.S.)	3.785 liters
1 gram	15.432 grains
1 kilogram	2.2046 pounds
1 metric ton	.9842 ton of 2240 pounds 19.68 cwts. 2204.6 pounds
1 grain	.0648 gram
1 ounce avoirdupois	28.35 grams
1 pound	.4536 kilograms
1 ton of 2240 lb.	1.1016 metric tons 1016 kilograms

FIGURE A2-128   Measures of volume and capacity.

Quantity	Unit	Symbol
Time	Second	s
Plane angle	Radius	rad
Force	Newton	N
Energy, work, quantity of heat	Joule	J
	Kilojoule	kJ
	Megajoule	MJ
Power, heat flow rate	Watt	W
	Kilowatt	kW
Pressure	Pascal	Pa
	Kilopascal	kPa
	Megapascal	MPa
Velocity, speed	Meter per second	m/s
	Kilometer per hour	km/h

FIGURE A2-129 Metric symbols.

Index